Dolls and Accessories 1910-1930

Dian Zillner

Schiffer Publishing Ltd

4880 Lower Valley Road, Atglen, PA 19310 USA

Dedication

This book is dedicated to the Denver Gals: Jan Hershey, Ellyn McCorkell, Paula Reding, and Nancy Roeder. These collectors furnished many pictures and/or allowed us to come into their homes to photograph dolls from their collections. Their help has made the series of doll books possible.

Notice: All of the items in this book are from private collections and museums. Grateful acknowledgment is made to the original producers of the materials photographed. The copyright has been identified for each item whenever possible. If any omission or incorrect information is found, please notify the author or publisher and it well be amended in any future edition of the book.

The prices listed in these captions are intended only as a guide, and should not be used to set prices for dolls and related products. Prices vary from one section of the country to another and also from dealer to dealer. The prices listed here are the best estimates the author can give at the time of publication but prices in the field can change quickly. This is especially true with the added sales from the Internet. Neither the author nor the publisher assumes responsibility for any losses that might be incurred as a result of consulting this price guide.

Title page photograph courtesy of Frasher's Doll Auctions, Inc., Oak Grove, Missouri.

Library of Congress Control Number: 2006931057

Designed by "Sue"
Type set in Grouch BT/Souvenir Lt BT

ISBN: 0-7643-2550-7
Printed in China

1 2 3 4

Published by Schiffer Publishing Ltd.
4880 Lower Valley Road
Atglen, PA 19310
Phone: (610) 593-1777; Fax: (610) 593-2002
E-mail: Info@schifferbooks.com

For the largest selection of fine reference books on this and related subjects, please visit our web site at **www.schifferbooks.com**
We are always looking for people to write books on new and related subjects. If you have an idea for a book please contact us at the above address.

This book may be purchased from the publisher.
Include $3.95 for shipping.
Please try your bookstore first.
You may write for a free catalog.

In Europe, Schiffer books are distributed by
Bushwood Books
6 Marksbury Ave.
Kew Gardens
Surrey TW9 4JF England
Phone: 44 (0) 20 8392-8585; Fax: 44 (0) 20 8392-9876
E-mail: info@bushwoodbooks.co.uk
Website: www.bushwoodbooks.co.uk
Free postage in the U.K., Europe; air mail at cost.

Contents

Introduction 5

Chapter One: Bisque Dolls 6
- Bisque Head Traditional Dolls 6
- All-Bisque Dolls 15
- Bisque Head Baby Dolls 24
- Bisque Head Character Dolls 33

Chapter Two: Celluloid Dolls 43
- Germany 43
- France 47
- Poland 47
- Japan 47
- United States 48

Chapter Three: Cloth Dolls 50
- United States 50
- England 60
- France 61
- Germany 61
- Italy 63
- Advertising Dolls 66

Chapter Four: Composition Dolls 68

Chapter Five: Metal Dolls and Heads 99
- Germany 99
- United States 104

Chapter Six: Wood Dolls 109

Chapter Seven: Comic and Celebrity Dolls 119
- Comic Dolls 119
- Celebrity Dolls 132

Chapter Eight: Boudoir and Pincushion (Half) Dolls 138
- Boudoir Dolls 138
- Pincushion (Half) Dolls 141

Chapter Nine: World War I Related Dolls 146

Chapter Ten: Furniture and Other Accessories 156

Bibliography 172

Index 174

A doll made with this china head was given to the author's mother, Flossie Ally, circa 1914. It was hoped that the doll would alleviate her disappointment when she discovered she was not old enough to begin school that year because of her December birthday. The doll is 16" tall with blonde molded hair, painted blue eyes with red eyelid lines and a closed mouth. The doll is marked "Germany" on the back shoulder plate. Her body and china limbs are replacements. Similar German dolls were sold for decades with either blonde or black painted hair. The Hertwig firm was still producing them through the 1920s.

Introduction

This is the third and final book in my series of doll books. The previous two are *Dolls and Accessories of the 1950s* and *Dolls and Accessories of the 1930s and 1940s*. While the 1950s dolls were mostly made of hard plastic or vinyl and the 1930s and 1940s dolls were nearly all composition, the period from 1910 to 1930 offered a large variety of dolls made from many different materials. When I began my doll collection in 1960, most collectors (including me) were concentrating on dolls from this time period. The first dolls I purchased included a Minerva metal head doll priced at $6.00, a carved head wood Schoenhut doll costing $10.00, a Chase cloth doll for $11.00, a Lenci doll also priced at $11.00, and a large bisque head doll on a ball-jointed body (with a fishing bobber replacing one of her ball joints) at the bargain price of $8.50. Since those days, I have changed direction and have accumulated a nice collection of both hard plastic and composition dolls. This book allowed me to return to my roots, so to speak, and revisit dolls from that earlier era.

The dolls from this period offer a wide variety of choices for the collector of today. They include bisque head dolls in infant or children styles (including character dolls) as well as all-bisque models. Also available are dolls made of celluloid, metal, composition, cloth, and wood. What an array of dolls to choose from!

In addition to the dolls listed above, this book also includes special chapters that picture and describe boudoir and pincushion dolls as well as the wonderful regular sized doll furniture from this period. It also includes a chapter on World War I related dolls, which describes how this war affected and changed the manufacture of dolls for years to come. Because of Germany's war preparations, the largest exporter of dolls in the world could no longer compete in this field. Manufacturers in the United States picked up the slack by producing dolls made of bisque, metal, composition, celluloid, cloth, and wood. Many of the companies formed to fill the void caused by Germany's inability to supply dolls to the world market did not survive, but several of the new firms continued to produce dolls for decades to come. The most successful of these American companies were Effanbee, Horsman, and Ideal. Some of the other doll manufacturers represented in the book include the German firms of Kestner, Armand Marseille, Kämmer & Reinhardt, Käthy Kruse, the French firms of Jumeau and S.F.B.J., and the Italian Lenci company. Other American firms include Louis Amberg & Son, Chase Stockinet, A. Schoenhut, and the Averill Manufacturing Co. Dolls from many other companies are also pictured and described.

The price of a doll depends on the condition, popularity, and rarity of the doll. The prices in this book reflect the dolls as they are pictured. If clothing has been replaced, this information will be included in the caption and the price given will be for a doll with replaced clothing. If the doll pictured includes the box, the price total will reflect that extra addition. No price is listed for dolls pictured only in catalog advertisements in order not to confuse the reader about the condition of the doll (box or no box, tag or no tag, etc.). Given the wide variety of dolls pictured in this book, it is difficult to provide an accurate idea of trends taking place in the doll world for all of them. It is safe to say, however, that any doll mint-in-box is very desirable no matter who produced the item or what material was used in its manufacture. Rare dolls follow the same criteria. Prices are also determined by the popularity and condition of a doll. Collectors seem to be especially interested in bisque character dolls, large 36"-42" bisque head dolls, googly dolls, small, so-called cabinet dolls (including bisque head 8" ball-jointed dolls and all-bisque Kestner dolls), mint Lenci and other rare cloth dolls, smoking boudoir dolls, and carved hair Schoenhut dolls in excellent condition.

I wish to thank the many collectors who have been so helpful in sharing materials, photographs, dolls, and knowledge with me so that this book could become a reality. These include: ARW Antiques of Saddle River, New Jersey; Susie Banchefsky; Ronald S. Barlow; Cobb's Doll Auction, Columbus, Ohio; Gail and Ray Carey; Patty Cooper; Denver Doll Emporium, Denver, Colorado; Frasher's Doll Auctions, Oak Grove, Missouri; Jean Grout; Jan Hershey; Nancy Lazenby; Ann Meehan; Leora Moberly; Marilyn Pittman; Carol Stover; Susie's Museum of Childhood at Bluebird Farm, Carrollton, Ohio; and Mary Lu Trowbridge.

I especially want to thank Paula Reding and Ellyn McCorkell, who allowed us to come into their homes to photograph dolls from their collections. A special vote of thanks also goes to Marge Meisinger, who, once again, loaned me catalogs from her fabulous collection so advertisements from the era could be photographed. She also supplied pictures from her great doll collection.

I would also like to extend continued appreciation and thanks to my daughter, Suzanne Silverthorn, who again took most of the photographs for this, my fourteenth book. A special "thank you" is also given to my son, Jeff Zillner, who helped with editing once again.

Acknowledgment and recognition is also extended to Schiffer Publishing Ltd. and its excellent staff, particularly to designer Sue Taylor and editor Donna Baker, who helped with this publication. Without their support and extra effort, this book would not have been possible.

Chapter One

Bisque Dolls

Bisque Head Traditional Dolls

In 1910, the doll industry had changed very little since the late nineteenth century. The German companies still reined supreme by exporting most of the world's supply of bisque, metal, and celluloid dolls. Although many of the French bisque dolls were superior to those made in Germany, they were more costly, which allowed the German dolls to retain their "best sellers" status. Some of the German firms had begun to experiment with bisque baby dolls and bisque character dolls before production shut down because of World War I. (See the "Bisque Head Baby Dolls" and "Bisque Head Character Dolls" sections later in this chapter.)

As the world situation began to change leading up to the war's beginning, German doll exports were curtailed. This allowed the United States to expand its own manufacture of dolls, even including those made of bisque. (See chapter on "World War I Related Dolls.")

The dolls pictured in this section are the more traditional dolls, primarily German, which had been made for decades. Types of bisque head dolls include those with shoulder heads, socket heads, bonnet heads, those with molded hair or wigs, dolls with painted eyes or glass eyes, and dolls that feature bodies of kid, cloth, papier mâché, or composition (including ball-jointed).

Dolls shown from German firms include those made by C.M. Bergmann, Heinrich Handwerck, Hertwig & Co., Ernest Heubach, Kestner, Kley & Hahn, Armand Marseille, and Recknagel. French dolls include a Bébé Jumeau and a #60 made by Unis.

Most of these dolls were made before World War I, but since many bisque dolls were produced from the same molds for decades, some were probably sold after the war was over, when production resumed in the 1920s. Several models, including the small slim dolls, the Unis #60, and the tiny Scotch boy, all date from the 1920s.

C.M. Bergmann

C.M. Bergmann was a German doll company that was founded in the late 1800s and often used bisque heads made by other firms. These included heads made by Armand Marseille and Simon & Halbig. The firm marketed baby, toddler, and child dolls.

28" tall bisque socket head doll on composition jointed body. She has blue sleep eyes, an open mouth with teeth, and a replaced wig. She is marked on the back of her head "Made in Germany/C.M. Bergmann/A 11 M." She has been redressed in an antique toddler dress. ($450-500).

Heinrich Handwerck

The Heinrich Handwerck German doll factory was located in Waltershausen, Thuringia, Germany. This firm used heads made by the Simon & Halbig firm. One of their most popular dolls was the "Daisy" doll, promoted by the *Ladies Home Journal* magazine in 1911 as a premium doll. She was 18" tall and had a bisque socket head with a composition ball-jointed body. The doll could be obtained for three subscriptions to the magazine. The firm made their own composition ball-jointed bodies.

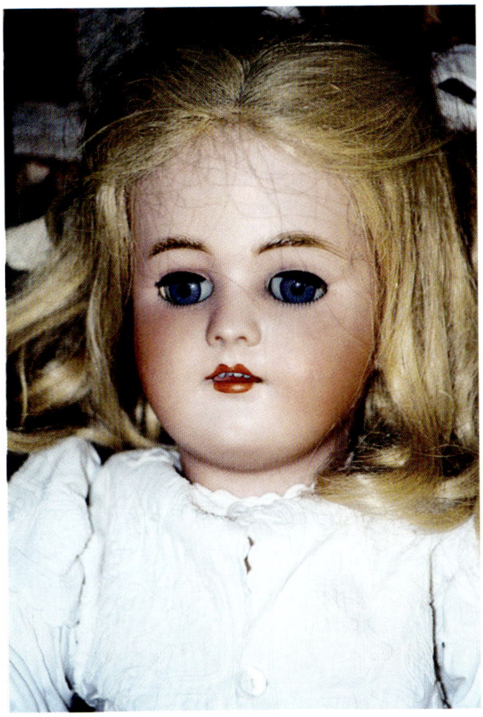

21" tall bisque socket head doll marked on the back of her head "Simon & Halbig /Germany/Heinrich Handwerck/ 3." The body is stamped Heinrich Handwerck on the lower back. The head was made by Simon & Halbig but the doll was completed and issued by Heinrich Handwerck. She has sleep eyes, open mouth with teeth, pierced ears, her original wig, and a ball-jointed composition body. She has been redressed. ($450-500). *Doll from the collection of Jeff Zillner*

Hertwig & Co.

Hertwig & Co. was a German porcelain factory that was in business from 1864 until 1953. The firm was known for its china head dolls beginning in the nineteenth century and continuing into the twentieth century. One line of these low brow dolls included different names that were marked on the front of the dolls. Bonnet head dolls made with untinted bisque were also popular items produced by the firm circa 1900-1915. Another favorite Hertwig product was the family of dolls pictured in their 1914 catalog. The dolls came in various sizes and included an older couple, a young woman, a young man, and a boy and girl. These dolls had bisque heads, upper bodies, hands, and lower legs, with cloth bodies, upper arms, and legs. As noted, they were featured in the 1914 Hertwig catalog. Many pages from this catalog are pictured in the book *Hertwig & Co. Archives 1890-1937*, by Florence Theriault. The catalog also pictures several bonnet head dolls, all-bisque dolls (including some with molded clothing), as well as many other Hertwig products. The firm was taken over by the East German government in 1953.

8" tall white bisque bonnet head doll model #208 pictured in the Hertwig & Co. catalog in 1914. It came in thirteen sizes. The dolls were sold undressed. She has a white molded bisque bonnet trimmed in pink flowers and a blue bow. Her hair is molded and painted as are her features. Her lower arms and legs are white bisque with glazed brown boots. All of the Herwig bonnet dolls were made with shoulder heads. ($150+).

11" tall "Butterfly" white bisque bonnet head doll made by the German Hertwig firm. She wears a molded pink bonnet in the shape of a butterfly. This doll was also made in a variety of sizes with either a pink or blue bonnet. She has painted features, cloth body, lower white bisque arms and legs, and glazed boots. She has been redressed. She may be model #203, circa 1900-1910. ($200+).

12" tall Hertwig white bisque bonnet head doll model #206, circa 1900-1910. Many of Hertwig's bonnet head dolls were sold for decades so it is hard to date a doll to a certain year. According to the *Hertwig & Co. Archives* book by Florence Theriault, this doll was featured in the firm's 1900 era catalog. She also came in thirteen sizes. The doll wears a molded green ruffled bonnet trimmed with an orange bow. Her blonde hair is molded and her features are painted. The shoulder head is decorated with another orange bow and a molded shirtwaist. She has a cloth body, lower white bisque arms and legs, and glazed brown boots. She has been redressed. ($200+).

9" tall Herwig doll wearing a bisque molded tam. She appears in the 1914 catalog as model #355. Her unusual red muslin body is original. She wears a white molded tam with a blue top-knot, molded blonde hair, and painted features. Her shoulder head includes a molded white ruffle edging. The same molded white ruffle trim is repeated above her bisque hands and her bisque molded stockings. Her shoes are yellow. She has been redressed. Some glue has stained her check. She came in eight sizes in the catalog. ($50+).

10" tall "snow baby" doll attributed to Hertwig. She is wearing a molded winter cap and has been dressed in a winter outfit. Although most snow babies are made of all bisque, this doll has a cloth body with lower bisque arms and legs. Her features are painted ($250-300). *Photograph and doll from the collection of Jean M.Grout.*

6.5" tall Hertwig doll with a one piece head and upper body made of molded bisque. This doll is one of a set of four different child character dolls featured in the 1914 Hertwig 50th Anniversary catalog. The other girl had short molded hair decorated with a molded hair bow. One of the boys featured a hairstyle with bangs and wore a molded bow at his collar. The other boy's molded clothing included an open red vest. The dolls came in ten sizes. The boy dolls wore cloth black pants and the girls wore plain brown skirts with cheaply made aprons. Their regular cloth brown arms provided the clothing for their arms. All of the dolls were dressed very cheaply. The example pictured here has been redressed. She has a molded braided hairstyle, painted features, and a molded vest and blouse front. Her hands are bisque as are her yellow shoes and molded stockings. She has been redressed. She is #384. Father and mother dolls in this style were also produced by Hertwig. They were featured on the front of the 1914 catalog along with the children as shown in the Theriault book. ($150).

5.5" tall shoulder head bisque doll also pictured in the Hertwig 1914 catalog. This model was #354. She was listed in eight sizes and in the larger sizes she had the red muslin body and parts exactly like the larger doll with the tam. This smaller example has the more usual brown sawdust filled body. She has blue molded hair bows, molded blonde hair, painted features, and a molded ruffle around her shoulders. Her hands, blue shoes, and socks are molded bisque. She has been redressed. ($75-100).

8" tall Hertwig shoulder head, sold in the 1914 company catalog as either a head alone or made up as a complete doll. The heads came in thirteen sizes while the complete dolls were offered in sixteen sizes. The bisque shoulder heads had molded light brown hair, eyes looking to the side, and closed mouths. The back of the shoulder is marked "Germany." This doll has a replaced body and reproduction parts. She has been redressed. ($50).

Ernst Heubach

The Ernst Heubach factory was located in Koppelsdorf, Thuringia, Germany. The dolls are usually marked Heubach-Kopplesdorf or with a horseshoe and/or the initials E.H. This firm's dolls are not as desirable as those made by the Gebrüder Heubach company. They did produce a variety of bisque head dolls including those representing children, babies, and toddlers. The dolls were made with either kid or composition bodies including ball-jointed styles.

Bisque socket head, composition ball-jointed dolls made by the German Ernst Heubach firm. The doll dressed in pink is 12" tall, has sleep eyes, an open mouth with teeth, and a replaced wig. She is marked on the back of her head "Heubach Koppelsdorf/312 SUR (in circle) 10/0 Germany." The doll in the rocking chair is also 12" tall. She has sleep eyes, an open mouth with teeth, and a mohair wig. Her visible marking reads "Koppelsdorf/Germany." Her hair covers the rest of her marks. She sits in a rocker from the period. Both dolls have been redressed. ($100-125 each).

21.5" tall bisque shoulder head doll with blue glass sleep eyes, open mouth with five teeth, and a brown mohair wig. Her kid body is jointed with metal pins at the shoulders, hips, and knees. The lower arms are bisque. She is marked with a horseshoe design on the back of her shoulder plate along with "Made in Germany." This mark was sometimes used by the Ernst Heubach firm. She has been redressed using a dress worn by the owner when she was a child. ($175+). *Photograph and doll from the collection of Marilyn Pittman.*

Jumeau

The French Jumeau Bébé dolls were sold from the 1880s until at least the 1920s. The earliest dolls had closed mouths while later dolls featured open mouths with teeth. The factory made chemises for the dolls were very similar in style from 1892 until the 1920s or later. The example pictured is dressed in this style original chemise. From 1900 on, the dolls were sold through S.F.B.J. The French Jumeau firm also manufactured fashion lady dolls, child dolls, character dolls, and a Princess Elizabeth doll (1938), as well as other models very much in demand by collectors.

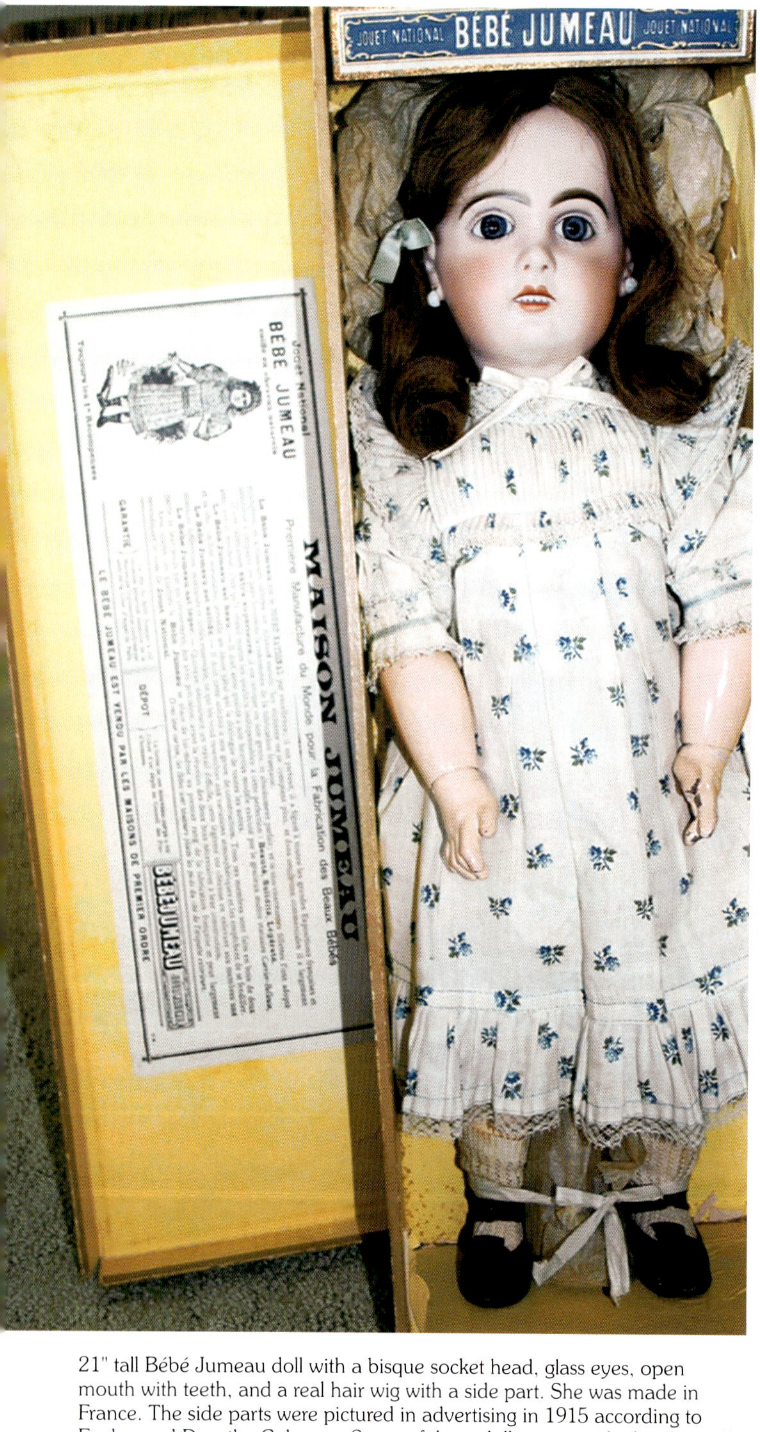

21" tall Bébé Jumeau doll with a bisque socket head, glass eyes, open mouth with teeth, and a real hair wig with a side part. She was made in France. The side parts were pictured in advertising in 1915 according to Evelyn and Dorothy Coleman. Some of these dolls were marked with a red stamp that read "Dépose/Bébé Jumeau." She wears her original chemise and shoes and socks. The doll has a fully jointed composition body. ($2,500-3,000). *Photographs and doll courtesy of Paula Reding and the Denver Doll Emporium, Denver, Colorado.*

This close-up of the Bébé Jumeau shows her pierced ears with earrings and the many tucks in her chemise. These dolls wore the same style chemise from 1892 until 1923 and perhaps later. The material came in two styles with either red flowers or blue flowers on a white background. Some of the dolls included belts marked "BÉBÉ JUMEAU." This doll's box is marked "Jouet National BÉBÉ JUMEAU." *Courtesy of Denver Doll Emporium.*

The end of the box is labeled "BEBE JUMEAU/Chemise Yeux 9 Coiffé." *Denver Doll Emporium.*

Kestner

The J.D. Kestner firm produced many different models of quality bisque dolls through their many years in business. Most of these dolls were made from the 1880s until the company closed in the 1930s. (See sections on "Baby Dolls," "Character Dolls," and "All-Bisque Dolls" for more examples of Kestner products.) The firm's traditional bisque head dolls were made in both shoulder head and socket head styles and included kid or composition bodies. One of the dolls pictured is model number 154, popular with both early consumers and current collectors. This model number was sold for decades in both the socket head and shoulder head styles. The other pictured example is model number 147, a lesser known Kestner doll. Both dolls have open mouths. Other Kestner mold numbers for these types of dolls include 129, 132, 133, 134, 141, 142, 143, 144, 145, 146, 148, 149, 152, 155, 156, 159, 160, 161, 164, 166, 167, 168, 170, 171 (Daisy), 173, 174, 192, 195, 196, 214, 215, 250, and 264, as verified by author Jan Foulke in her book *Kestner: King of Dollmakers*.

Bisque shoulder head dolls made by the German J.D. Kestner firm. The smaller doll measures 13.5" tall and has brown stationary glass eyes, an open mouth with teeth, and a real hair wig. She has a jointed kid body with lower bisque arms. She is marked on the back "Dep 154" and 1 ½". The larger doll is 16.5" tall and has blue glass sleep eyes, an open mouth and teeth, and a mohair wig. The shoulder head is mounted on a jointed kid body with lower bisque arms. The back of the head is marked "147". Both dolls have been redressed. (Smaller $150, larger $250). *Dolls from the collection of Jeff Zillner.*

Armand Marseille

The German Armand Marseille firm was one of the most prolific of the bisque doll manufacturers. (See also the bisque "Baby Dolls" and "Character Dolls" sections.) Besides producing dolls under their own name, the firm manufactured bisque doll heads for other companies as well. Traditional type dolls from A.M. included Floradora, Queen Louise, named dolls like Alma and Mabel, and their very popular model #390. Most of these examples were made in both shoulder head and socket head styles with either kid or composition bodies. The firm's 8" socket head dolls mounted on a ball-jointed body are particularly appealing.

28" tall Queen Louise doll by the German Armand Marseille firm, circa 1910. She has a bisque socket head, glass sleep eyes, an open mouth with teeth, her original mohair wig, and a composition ball-jointed body. She is marked on the back of the head "Queen Louise/Germany." The doll is wearing what appears to be her original clothing, although it is now in tatters. ($350-400). *Doll from the collection of Jeff Zillner.*

33" tall large child doll made by Armand Marseille. She has a bisque socket head, composition ball-jointed body, stationary glass eyes, and an open mouth with teeth. Her real hair wig appears to be original. She is marked on the back of her head "Armand Marseille/ Germany/390/A.15M." She has been redressed in a child's dress circa 1910-1915. The #390 model was a very popular doll and it sold in a large variety of sizes. Shoulder heads were also produced with this same model number. ($600+).

7.5" tall "topsy turvy" doll made by Armand Marseille. She has two bisque shoulder heads (one white and one black). The lighter head has glass eyes, an open mouth and teeth, and a mohair wig. Her lower arms are made of composition. The back of this shoulder head is marked "3200/AM12/0dep/Germany." The dark head has painted features, molded hair, and lower arms of bisque. A cloth body connects the two shoulder heads. The dress for the dolls is original and includes a long reversible skirt that can be used with either head to make a white or black doll. The skirt covers the doll not in use. ($250+).

8" tall bisque socket head Armand Marseille doll with a ball-jointed body. She is jointed at the neck, shoulders, elbows, hips, and knees. She has sleep eyes, an open mouth with teeth, and her original mohair wig. She is wearing one of the original handmade dresses that were part of the wardrobe made for her when she was new. The wardrobe included a coat and hat, rompers, and several additional dresses. She is marked "A.M." along with "Dep" and some numbers that can't be read. ($250-300).

7" tall bisque socket head doll with a five-piece composition body, circa 1920s. This Armand Marseille doll has sleep eyes, open mouth with teeth, her original bobbed mohair wig, and molded shoes and stockings. She is wearing her original clothing including a large hair ribbon. ($175+). *Doll from the collection of Ellyn McCorkell.*

A similar doll boxed with her wardrobe was advertised in the Sears 1923 Fall and Winter Catalog. The 7" tall doll came with twenty-three pieces, including a dress, hat, jacket, cape, romper dress, comb and mirror, bathing cap and bag, towel, imitation soap, and sand toys. The set came in an attractive box and sold for $2.48 complete.

Recknagel

The German Recknagel firm produced dolls of varying quality. Their babies with molded bonnets are especially attractive. (See section on "Baby Dolls.") Most of the bodies on their dolls, however, are inferior in quality. The dolls are still very collectible, especially the babies, googly-eyed, and the Max and Moritz dolls. Most of these dolls were produced from circa 1890s until World War I.

9" and 8" bisque socket-head dolls made by the German Recknagel firm. The larger doll has glass sleep eyes, an open mouth with teeth, and a mohair wig. Her arms and legs are made of papier mâché with molded and painted shoes and socks. She is wire jointed at the shoulders and hips. The body is made of cardboard. She wears her original clothing. The head is marked "RA." The smaller doll has glass eyes, open mouth and teeth, and a mohair wig. Her arms and legs are crude, made of papier mâché and wire jointed at the shoulders and hips. Her body is also made of cardboard. Her hair ribbon and dress are original. The back of her head is marked "DEP/R15/OA." ($75-100 each).

Unis France

The Unis France mark was used by S.F.B.J. circa 1916-1930+ (see page 42). Some of their model numbers included #60, 301, and with toddler bodies #247 and 251.

12" tall brown bisque Unis socket head doll circa late 1920s, early 1930s. She has brown glass eyes, open mouth with teeth, and a mohair wig. Her brown body is composition and is jointed at the shoulders and hips. She is all original. She was made in France and is marked "Unis/France/60." ($250).

Unidentified

Many of the traditional bisque head dolls have been identified as to maker through research, original boxes, and advertisements. However, there are still a variety of dolls that include words, initials, or number markings and have not yet been matched with the companies that produced them. The following dolls are included as representations of this category.

15" tall bisque shoulder head doll on a kid body. She has stationary eyes, open mouth and teeth, and lower bisque arms. Her wig has been replaced. She is marked "Special" in script as well as "9/0." The German Adolf Wislizenus firm sometimes used the "special" label on bisque heads. ($150+). *Doll from the collection of Jeff Zillner.*

9" tall bisque tinted brown socket head Indian doll made in Germany. The doll is marked in a triangle on the back of the head with letters that look like "C F & S." She has brown glass eyes, an open mouth with teeth, and a mohair wig. Her light brown body is a nicely made five-piece composition one with molded shoes. The Indian felt costume is all original. She has the slightly wavy eyebrows seen on Indian dolls from several companies. ($150+).

5.5" tall bisque socket head doll dressed in Scottish costume. He is marked with a symbol on the back of his head but since the head is so small, it can't be identified. He has tiny stationary glass eyes, closed mouth, and a small bit of mohair showing under his cap. The composition body is nicely made and is jointed at the shoulders and hips. He has molded and painted shoes. The doll is all original and finely dressed. Perhaps he was originally sold as a souvenir doll in Scotland circa 1920s or 1930s. ($75+).

6" and 5" bisque socket head dolls circa 1914. Marshall Field featured these German dolls in their "Kringle Society Dolls" catalog in 1914. Twenty-two different dolls were illustrated on the page, including these two examples. The German dolls ranged in size from 5" to 6.5" high. Many of the dolls had glass eyes. The larger doll pictured was called a "Nurse with Baby" in the catalog. She has painted eyes, closed mouth, and a mohair wig. Her five-piece composition body is of medium quality. Her shoes are painted. The clothing is all original and includes a tiny, all bisque "Frozen Charlotte" baby tied to her arm. The boy doll also has painted eyes, a closed mouth, and a five-piece composition body. His sailor suit is all original and is nailed to his body with a tiny nail in the back. The hat is also original. (Pair $200).

5.5" tall bisque socket head doll marked on the back of her head "S IV." This doll is an earlier and cheaper example of the small bisque socket head doll. She has painted eyes, closed mouth, and a mohair wig. Her original clothing is nailed to her body. The head is mounted on a crude five-piece papier mâché type body with molded and painted shoes. She is probably circa 1910. ($75+).

5.5" tall boy and girl "twin" dolls are also similar to several pairs offered in the Marshall Field 1914 catalog. These dolls have painted eyes, closed mouths, mohair wigs, and five-piece composition bodies. Their shoes and stockings are molded and painted. They wear their original sailor outfits. (Pair #200+). *Dolls from the collection of Ellyn McCorkell.*

All-Bisque Dolls

With the coming of the Internet, doll prices have changed dramatically. The common dolls have continued to lose value while the prices of the unusual and hard to find examples have generally continued to hold their own and some have escalated in price. This is especially true of the all-bisque dolls. New research has made sellers and collectors more knowledgeable and it is no longer possible to find a swivel neck all-bisque Kestner example at a bargain price. The increase in value for these dolls has also been affected by their convenient size. A wonderful varied collection of these dolls can be displayed in a small cabinet that can follow the owner into an apartment or condo later in life.

Some of the most collectible (and expensive) of these dolls are the so called French styled dolls. Most of these dolls were made before 1910, beginning around 1880. Genevieve Angione has some wonderful examples of these dolls in her book *All-Bisque and Half-Bisque Dolls*. For the most part, the dolls have a slimmer head and body than the more plentiful German all-bisque dolls. Most of them have glass eyes, swivel necks, and are peg strung. Fewer examples are found with ball joints made of bisque. These joints have been found at both the elbows and knees. Other dolls have peg strung elbow joints. Many of these early dolls include molded bare feet. Other identification markers are upper and lower painted eyelashes, one stroke eyebrows that almost meet, and mold seams that are sometimes off center or down the center instead of at the sides of the dolls. For those collectors who are interested in these earlier dolls, the Angione book is recommended.

For collectors of all-bisque dolls produced from 1910-1930, there are many fine examples to choose from. The character dolls are currently very popular. They include dolls based on comics like Buster Brown, Little Orphan Annie, Little Annie Rooney (really based on Mary Pickford's movie of the same name), Max and Moritz (with molded clothes), and the other comic characters pictured in this book's comic chapter.

Other all-bisque character dolls were based on illustrations famous during the years. They include Kewpies and Scootles (both Rose O'Neil designs), HEbee SHEbee, (Charles Twelvetrees), and September Morn (a doll designed by Grace Drayton).

Some of the all-bisque character dolls were developed because of the popularity of larger dolls made in the same image. These include Baby Peggy Montgomery, made in Germany and marketed in several sizes (3"-4.5") by Louis Amberg & Co. This doll was produced with either molded hair or in a wigged style. The successful Mibs doll was also marketed by the Amberg firm in an all-bisque form, ranging in size from 3"-6" tall. Joseph L. Kallus' Baby Bo Kaye produced by the German firm of Alt Beck & Gottschalck was another successful large doll that was also produced in miniature all-bisque form. He was 6" tall and had molded hair and glass sleep eyes. Two of the most popular baby dolls of the period, Bonnie Babe (designed by Georgene Averill) and the Bye-Lo Baby (designed by Grace Storey Putnam), are now nearly as desirable for collectors in their small sizes (Babe: 5", 7") as they are in their original larger sizes. The small Bye-Los were manufactured in a variety of styles, including painted or glass eyes, painted hair or wigs, with bare feet or molded booties, and jointed or stationary necks. Sizes for the small Bye-Los ranged from 4" to 8" tall. E.I. Horsman even marketed a small, all-bisque Tynie Baby doll based on their larger best seller. The smaller version came with painted or glass eyes.

All-bisque character dolls were additionally produced to represent the Happifats characters designed by Kate Jordan. These dolls were made in both Germany and Japan. The Teenie Weenie characters designed by W. Donahey also became miniature all-bisque dolls circa 1922. Chinese characters were found to be good sellers when produced in doll form during the 1920s. The German G. Heubach firm manufactured a small all-bisque doll called Chin Chin and the similar Queue San Baby was probably made in Japan.

In addition to the already mentioned dolls, the all-bisque dolls designed by Jeanne I. Orsini from 1916-1920 are becoming even more expensive than the early French dolls. Produced by the German firm of Alt, Beck, and Gottschalck for George Borgfeldt, they range in size from 5" to 7" tall. Orsini designed her dolls in a studio located in New York. Her first copyright was for a doll with a crying face called "Tummyache" in 1916. Some of the later dolls came with a paper label that gave the name of the doll plus "REG. U.S. PAT.OFF./COPR. J.I. ORSINI/1920 PAT. APPL'D FOR/GERMANY." Most of the 1920 designs were incised on the back JIO c 1920. The 1919-1920 dolls were made in several different designs including Didi, Mimi, and Vivi.

The Orsini dolls had glass eyes and wigs and were jointed at the shoulders and hips. They also came with molded one-strap shoes and painted stockings that extended above the knees. The dolls had unusual character faces molded into different expressions including wide smiles.

All-bisque dolls can be divided into a number of different categories, organized by maker (problem, very few were marked), by type, by those with molded clothing and/or hats versus those meant to wear regular clothes, by chronological order according to date or origin, or by children as opposed to babies. A combination of these methods will be used in this chapter.

Although many catalogs are available that include advertisements for dolls from the early part of the twentieth century, most of these sources do not picture all-bisque dolls because they were sold so cheaply. The firms that did handle such dolls usually pictured only drawings of the dolls, making it generally difficult to identify a particular doll. *The Marshall Field & Co. 1914 Kringle Society Dolls'* catalog, however (reprinted by Hobby House Press in 1980) provides a wonderful source for studying the all-bisque dolls for sale at this time because the catalog used photographs instead of drawings to represent the dolls. In three pages covering all-bisque dolls, three dolls wearing molded clothing are shown (Indian, boy in sailor suit, and girl in coat and hair bow), two dolls with stationary legs are pictured, along with three child dolls with molded hair. In addition five all-bisque baby dolls are shown (one with wig), and twenty-four wigged little girls (one labeled Negro) are featured. The girls with glass eyes number twelve (including the Negro). Some of these dolls have sleep eyes, and others have stationary eyes. The prices are given in wholesale numbers with the cheapest dolls being the dolls with molded hair (dozen 80 cents), some of the dolls with wigs and painted eyes, as well as the smaller babies. The most expensive dolls sold for $8.50 a dozen in 8" and 9" sizes. The most expensive babies had painted eyes and were 10" tall. These sold for $16.00 a dozen. All the dolls had either wire jointed limbs or were jointed with rubber cord. The shoes varied from having no straps, to one strap, to two straps. Many of the dolls had shirred stockings. The wigs on the dolls (all girls) included very long hair, except for one example in which the hair was braided over both ears. About half of the wigs included bangs and the rest were parted in the middle or on the side. All of the girl dolls were sold nude and were molded in the chunky style of the German manufacturers of the period.

By the late 1920s, most German firms were producing their small dolls with bisque heads and composition type slim bodies to keep up with the new 1920s Flapper look. The cheaper painted all-bisque dolls in slim sizes were made in both Germany and Japan and filled the inexpensive small doll market beginning in the 1930s. These dolls had painted molded hair and painted features and sold mostly in "dime stores" and similar outlets.

Molded Clothing and/or Hats

Although most of the all-bisque dolls were made so they could wear real clothing that could be removed, some of these dolls wore molded clothes. Many had moving arms only but some included both jointed arms and legs. Others were nodders or had no moving parts at all. (See Comic chapter.) Both German and Japanese firms produced these types of dolls. The German firms of Hertwig and Co., for example, made many different kinds of these types of dolls. Since most of the dolls are not marked, it is difficult to know which firm was responsible for each product. Most of the Japanese dolls are marked "Nippon" or "Made in Japan."

Besides the dolls pictured, other dolls wearing molded clothing include the following models made by Hertwig: Two examples of boys in sailor suits (one with hat and jointed legs, one without hat and stationary legs), boy in rompers and boots, boy in rompers and shoes, Indian Boy (moving arms), cowboy and cowgirl (moving arms), boy in sailor suit and hat and girl in coat dress and hair bow (pictured in 1914 Marshall Field catalog, with turning heads and stiff limbs), girl in Chinese-looking pajamas (jointed arms), and snowbabies (jointed arms and legs). Other known makers of all-bisque dolls with molded clothing include Bonnie Babe by Georgene Averill, Max and Moritz by Kestner, Chubby (S.H.K. Co., also seen with label L.W. & C for Louis Wolf & Co.), and Happifats designed by Kate Jordan.

Additional dolls with molded clothing not yet identified as to maker include a baby with molded long shirt, Dutch girl with hat, girl in hat, bathing suit and swimming shoes, and many more.

All-bisque dolls with molded hats only were also produced in Germany for decades. The earlier dolls have very fancy hats or bonnets. These dolls included both moving arms and legs or only jointed arms. The later molded hats were tams or the 1920s cloche style hats.

5.5" and 7" tall all-bisque boy dolls made by the German firm of Hertwig & Co. Both dolls wear molded sweaters, pants, shoes, socks, and caps. The smaller doll has wire jointed arms and legs while the larger doll has only wire jointed arms. A smaller version of the larger doll appears in the 1914 Hertwig & Co. catalog. The hair is molded and the features are painted on both dolls. The larger doll has red lines painted over the eyes and both dolls include red dots to indicate the nostrils. ($150+ each).

4" tall all-bisque boy with molded clothing, made in Germany. He has painted hair, big side glancing eyes, and large feet. His arms and legs are stationary. He looks very much like the "Little Jimmy" character drawn by James Swinnerton in the 1920s. He is stamped "Made in Germany" on the bottom of his shoe. ($50+). *Doll from the collection of Marge Meisinger. Photograph by Carol Stover.*

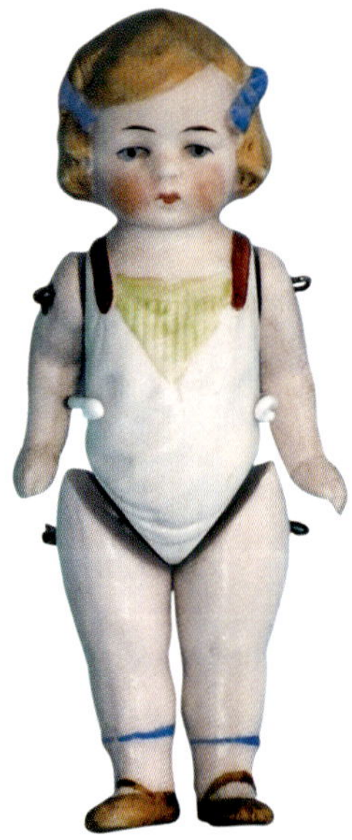

4.5" tall all-bisque little girl, circa late 1920s or early 1930s. She has molded hair featuring two blue hair bows. She is wearing a molded bathing suit or sun suit. The doll is wire jointed at the shoulders and hips. Although she is not marked, she appears to have been made in Germany. ($50-75).

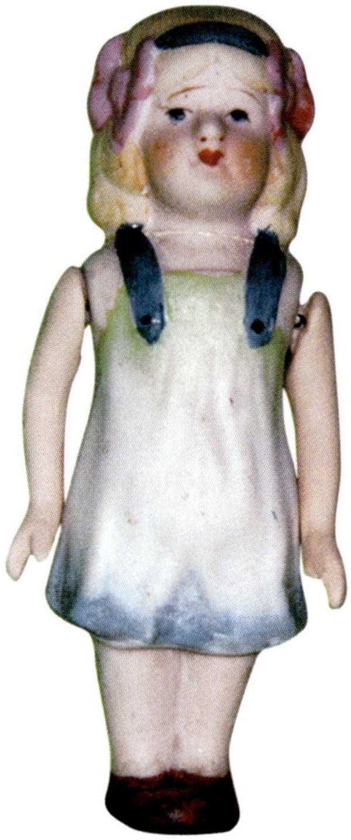

5" tall all-bisque little girl with jointed arms, circa 1920s. She has molded hair, headband with bows, and a molded dress with straps. She is marked "Nippon" on the back to indicate that she was made in Japan. ($35-50). *Doll and photograph from the collection of Marilyn Pittman (her mother's childhood doll).*

5" tall all-bisque doll marked "3/0" plus something unreadable on her back. She appears to be a German product from the 1920s. She wears a molded hat with a 1920s look, has eyes looking to the side, and a watermelon shaped mouth. She is jointed at the shoulders and hips and wears black painted one-strap shoes and shirred stockings. She has been redressed. ($125+).

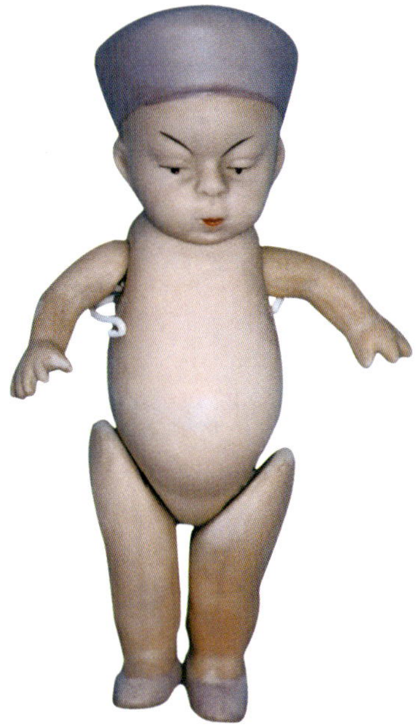

5" tall all-bisque Queue San Baby. He wears a molded hat with a molded black pigtail down his back. The doll is cord strung. According to information in Genevieve Angione's book, *All Bisque and Half Bisque Dolls*, the first Queue San Baby was designed by Hikozo Arakia, a U. S. citizen. The patent was assigned to Morimura Bros. in 1916. Several different dolls with this molded hat, face, and pigtail have been produced, including a kneeling figure with jointed arms. The dolls have starfish hands. This example has jointed arms and legs. ($75-100).

4" tall Happifats girl and boy all-bisque dolls designed by Kate Jordan and patented in 1914. The dolls were made in both Germany and Japan. The back of the little girl is marked "Nippon." Both are wearing molded clothing, including a pink dress trimmed with a blue sash (girl) and brown pants, green jacket, and a white shirt (boy). Both dolls have jointed arms. The Happifats made in Germany include a little girl wearing a blue dress with a pink sash. A baby with a molded diaper was also produced. It, too, included jointed arms. The Butler Bros. catalog of 1914 pictured a Happifats set that cost 39 cents a pair wholesale. ($250 pair). *Dolls from the collection of Ellyn McCorkell.*

Left:
5" all-bisque doll wearing a molded hat that has the look of a Japanese product but is not marked. The hat appears to be from the 1920s. She has jointed arms and stationary legs. Although her head and hat have been colored, her body and arms remain white. This is the type of doll that sold cheaply in the dime stores for many years. ($40+).

Molded Hair

6" tall all-bisque character girl with mold number 10490 made by the German Gebrüder Heubach firm. She has brown molded hair decorated with three molded pink bows. The doll has painted features with an open/closed mouth and painted shoes and socks. She is jointed at the shoulders and hips. She has been redressed. ($700-1,000). *Doll courtesy of Paula Reding and The Denver Doll Emporium.*

Unusual 4.25" tall all-bisque soldier doll, circa early 1900s to World War I era. He has molded hair, painted features, and is jointed at the shoulders and hips. The doll is wearing his original clothing including a metal helmet. Because of the placement of his hands, he may have originally been holding a drum or other object. ($150-175). *Photograph and doll from the collection of Gail and Ray Carey.*

4" and 5" tall all-bisque German boy dolls, both wire jointed at the shoulders and hips. Both have light brown molded hair and painted features. The smaller doll wears molded vertical shirred stockings while the larger doll's stockings are molded in a pattern of small squares. Both wear shoes with heels. (Small $75, large $100).

All-bisque molded hair dolls were made with both stationary and jointed legs. Pictured is a 6.5" tall all-bisque boy dressed in red with jointed arms and legs, and a 4.75" tall boy with jointed arms and stationary legs. Both are from Germany. The smaller doll is older, as shown by his two-strap bootines and shirred stockings. He has molded blonde hair, blue painted eyes, and closed mouth. The doll has been nicely redressed. The larger doll also has blonde hair, blue painted eyes, and a closed mouth. Although his stockings are shirred, his heeled shoes have only one strap. He has also been redressed. ($125-150 each).

Left:
6.25" tall all-bisque German doll dressed as a girl. She is jointed at the shoulders and hips and is very finely molded. She has light brown hair, painted blue eyes (with red eyeliner lines), and a closed mouth. Her hands include separated thumbs. She wears brown one-strap heeled shoes and stockings in the square shirred pattern. She is not marked and has been redressed in a dress made from a handkerchief. The doll could also be dressed as a boy. This model came in a 9" size as well. ($150+).

Japanese firms also produced all-bisque dolls of varying quality during the teens and 1920s. This 7.5" doll is wire jointed at the shoulders and hips and marked "Nippon" along with a company on the back. It may say "B.I. & Co. (B. Illfelder & Co.). The doll is like one owned by the author's mother during her childhood, circa World War I years. She has blonde molded hair, blue painted eyes with red eyeliner lines, a closed mouth, black one-strap shoes, and shirred white stockings. ($100-125).

Pair of all-bisque German dolls apparently used to decorate a wedding cake long ago. The "bride" still has her worn bridal veil. She is 4.5" tall and wire jointed at the shoulders and hips. The doll has molded blonde hair decorated with two molded blue hair bows, painted blue eyes, and a closed mouth. She is marked "Germany" on the back. Her heeled one-strap shoes and stockings (molded in a pattern of small squares) are like those on the previous 5" German boy doll. The 5.5" groom is also wire jointed at the shoulder and hips. He has molded blonde hair, painted blue eyes looking to the side, and a closed mouth. He wears brown one-strap shoes with no heels. ($100-125 each).

All-bisque German boy and girl dolls with similar shoes and socks. The little girl is 7.75" tall with molded light brown hair decorated with two molded blue hair bows. She is wire jointed at the shoulders and hips. She wears brown painted no heeled, one-strap shoes and short shirred stockings. She is marked on the back "P.85" in an oval, a three leaf clover design, and "Made in Germany." This marking indicates the doll was produced by the German Limbach firm. The 7" tall boy doll is also wire jointed at the shoulders and hips. He has molded blonde hair, painted blue eyes, and a closed mouth. He has no markings to indicate his maker. A similar 6.5" tall doll was reproduced by Ellery Thorpe a number of years ago. The reproduced doll has hands molded in fists, blue painted socks, and black shoes. This doll wears brown painted shoes and short shirred stockings. His body is pink tinted. Both dolls have been redressed. ($150-175 each).

6.5" and 7" tall all-bisque little girls with molded hair. The smaller doll is marked "Made in Japan." The larger doll is not marked but appears to be a Japanese product in quality. Both dolls have blonde hair, molded ribbons in their hair, painted blue eyes, and closed mouths. Both also are wearing painted black one-strap shoes and lightly vertical lined shirred stockings. ($75+ each).

7.5" tall all-bisque dolls probably made in Japan. This doll design was used for many years and was still being sold in the Butler Bros. catalog in 1932 for 72 cents a dozen in the 7.5" size. The dolls have molded hair, large eyes looking to the side, and jointed arms only. The pair has been redressed as a bride and groom and was perhaps used on a wedding cake at some point. ($150+ for pair). *Dolls from the collection of Marge Meisinger. Photograph by Carol Stover.*

J.D. Kestner

Probably the most collectible all-bisque dolls of the 1910-1930 period, in addition to the character dolls, are those that can be identified as being made by the German firm of J.D. Kestner. Their doll factory was located in Waltershausen, Thuringia, and the firm produced high quality bisque dolls from the 1880s until the 1930s.

One of the reasons these dolls are so collectible is that much research has been done to identify the mold numbers used by the firm on these small dolls. Jan Foulke's book, *Kestner: King of Dollmakers*, is particularly helpful in this regard. Some of the mold numbers used for Kestner all-bisque dolls include: 102, 130, 150, 160, 184, 190, 208, 310, 602, and 620. Many of the company's early all-bisque dolls wore the molded two-strap bootines on their feet. These dolls came in styles with stationary legs as well as jointed legs. Later, some dolls wore high gold boots like those pictured on "The Wrestler" with many straps. Later still, many Kestner all-bisque dolls wore the popular molded yellow boots now sought by collectors. By 1914, catalogs were picturing all-bisque dolls with two-strap heeled shoes as well as one-strap plainer footwear. Kestner used both styles on their dolls of the period.

Kestner all-bisque dolls were assembled in several different ways. The early dolls were put together with pegs at their shoulder and hip joints. Some of the parts are kid-lined in the earlier dolls. Later dolls are loop strung with elastic. Apparently Kestner did not use the wire strung method to assemble their dolls as did many other firms.

The most expensive all-bisque Kestner dolls are the googly models. The 189 mold came with either painted or glass eyes and with stationary or swivel necks. The model number 112 googly has jointed elbows and knees as well as jointed shoulder and hips. It has glass eyes, a fitted wig, and a watermelon painted mouth. Other expensive Kestner all-bisque dolls include model #102 usually called the "Wrestler."

In addition to the described dolls, Kestner also produced some all-bisque baby dolls. These are harder to identify as Kestner products unless they are marked or use the same head as a known Kestner baby doll that came on a composition body. The dolls were made from approximately 4.5" to 14" in size, according to Jan Foulke.

9" tall all-bisque so called "Wrestler" by Kestner. She is marked 102 and has a bisque swivel head on a five-piece jointed bisque body, stationary blue glass eyes, painted lashes, pierced ears, open mouth with teeth, original mohair wig, bent left arm, painted white socks, and gold six-strap bootines. Her outfit is pictured by her side. Although this doll has a repair to her neck, she is still a very desirable doll. (With repair $800-1,000). *Courtesy of Barbara Frasher and Frasher's Doll Auctions, Inc., Oak Grove, Missouri.*

5.75" tall all-bisque Kestner mold #130. She has sleep glass eyes and her original mohair wig. She is loop strung and jointed at the shoulders and hips. Her stockings are shirred in an unusual pattern and she wears heeled brown shoes with two straps, decorated with black bows on the toes. She has been nicely redressed. ($250-300).

9" tall all-bisque larger model of the Kestner #150. She has sleep glass eyes, a mohair wig, open mouth with teeth, and molded one-strap shoes and ribbed stockings. She has been beautifully redressed. ($700+). *Courtesy of Paula Reding and the Denver Doll Emporium.*

5.25" tall all-bisque Kestner mold #620. This doll is the same doll as the previous example except she has a swivel neck. She wears her original wig, has sleep glass eyes, and brown two-strap shoes decorated with painted black bows on the toes. She has been redressed. ($350-400).

5" tall all-bisque Kestner mold #150. She is probably a later example than #130. She wears the one-strap shoes and still retains her original Kestner sticker reading "Kestner/Germany." The doll wears her original braided wig (similar to a doll shown in the 1914 Marshall Field catalog with round braids on either side of her head). She has glass eyes and an open mouth with molded teeth. She is loop strung and jointed at the shoulders and hips. ($300+ with sticker). *Courtesy of Paula Reding and the Denver Doll Emporium.*

7" and 5" tall Kestner examples of the #602 mold described as Kestner products by Jan Foulke in her book, *Kestner: King of Dollmakers*. The all-bisque dolls have glass eyes, mohair wigs, and slightly open mouths with teeth. They are jointed at the shoulders and hips and also include swivel necks. Both examples are wearing black one-strap shoes and short molded stockings. The dolls have been redressed. (Small $350+, larger $500-600). *Courtesy of Paula Reding and the Denver Doll Emporium.*

All-Bisque Wigged Dolls

4.5" and 5" tall German all-bisque dolls made of tinted bisque. The smaller doll has facial features very similar to those pictured in the book, *Hertwig & Co. Archives*, by Florence Theriault. An original card of these dolls from Hertwig & Co. is pictured in the book. They range in size from 3" to 7" tall. All of the dolls have wigs. Most examples have painted eyes while a few have glass eyes. Nearly all of the dolls were wearing the same footwear as the two dolls pictured here. The Hertwig dolls, as well as the dolls pictured, were wire jointed at the shoulder and hips. The Hertwig dolls were circa 1914 and probably continued to be sold into the 1920s. The two pictured dolls have replaced wigs and have been redressed. The larger doll is marked on her back "Germany/2/0" along with 15 in pencil, which is assumed to be the original price. ($75-100 each).

3.75" tall all-bisque German little girl doll also wearing one-strap black painted shoes and shirred stockings. She has large brown glass eyes, closed mouth, and a wig that may be a replacement. She is jointed at the shoulders and hips and has been nicely dressed. ($125-150). *Doll and photograph from the collection of Marilyn Pittman.*

3.75" tall pair of German all-bisque dolls dressed in their original native costumes. They both have glass eyes, mohair wigs, molded and painted features, and are jointed at the shoulders and hips. The boy is wearing two-strap shoes while the girl wears the one-strap shoes. Both wear shirred stockings. The dolls are marked "36-10" on their backs. According to Jan Foulke writing in *Doll Reader* in October 1990, the mold number 36 indicates that the dolls were produced by the German porcelain factory of Kling & Co. Their products are sometimes marked with a bell. Since the all-bisque dolls were usually sold nude, it is hard to find older dolls in original outfits such as these. ($350-400 pair). *Dolls from the collection of Marge Meisinger, photograph by Carol Stover.*

3.5" and 3.75" (boy) tall all-bisque dolls made in Germany. Both dolls include their original mohair wigs, are wearing painted one-strap shoes with shirred stockings, and are jointed at the shoulders and hips. The boy has glass eyes while the girl's eyes are painted. (Boy $150, girl $100+). *Dolls and photograph from the collection of Gail and Ray Carey.*

3.75" tall pair of all-bisque dolls made in Germany. They have glass eyes, closed mouths, matching mohair wigs, and black shoes. Both dolls are jointed at the shoulders and hips. They are wearing matching sweaters, caps, and leggings. She wears a skirt while he wears matching pants. The little girl carries a doll while her brother holds a red metal ship. ($350-400 pair). *Dolls and photograph from the collection of Gail and Ray Carey.*

4" tall all-bisque doll with glass eyes, open and closed mouth, and mohair wig. She is jointed at the shoulders and hips. The doll wears two-strap heeled brown shoes and shirred stockings. She is marked "1/8/4 Germany," has been redressed, and may have a replaced wig. ($175+). *Doll from the collection of Marge Meisinger. Photograph by Carol Stover.*

4.25" tall all-bisque German doll. She has glass eyes, a mohair wig, closed mouth, and a swivel head. She is also jointed at the shoulders and hips, and has painted one-strap buckle shoes. She has been redressed. Her head is not marked although she looks very much like the Kestner mold #208. ($275-300). *Doll and photograph from the collection of Gail and Ray Carey.*

5" tall all-bisque Mimi doll designed by Jeanne I. Orsini in her New York studio. The dolls were produced in Germany circa 1920. Other Orsini dolls of the period include Zizi, Didi, and Vivi. The dolls had wigs, glass eyes, joints at the shoulders and hips, and very expressive faces. Didi was a 7" doll. Mimi has an open mouth with lips pursed as if to whistle. Her forefinger on the right hand is extended. She has a slender body with painted shoes and painted long stockings. She has blue glass eyes and a very long brown human hair wig. She is marked on the back of her shoulders "J 10 c.1920/44." ($2,200+). *Photograph and doll courtesy of ARW Antiques in Saddle River, New Jersey.*

All-Bisque Baby Dolls

2.25" all-bisque German baby doll with molded hair and painted features. He is jointed at the shoulders and hips. Most of these tiny dolls were displayed in filigree metal buggies or used in doll houses. The buggies were sold for many years. Similar soft metal buggies were advertised in the Butler Bros. catalog in 1914. (Set $75).

4.5" tall all-bisque German baby doll that includes a bottle molded in its hand and an open mouth to receive the bottle. The other hand has a separately molded thumb so it can also be placed in the doll's mouth. The baby is marked on the back "773/9 ½". The doll has lightly molded hair, small painted eyes, and a swivel neck. He is also jointed at the shoulders and hips and has curved arms and legs. ($100).

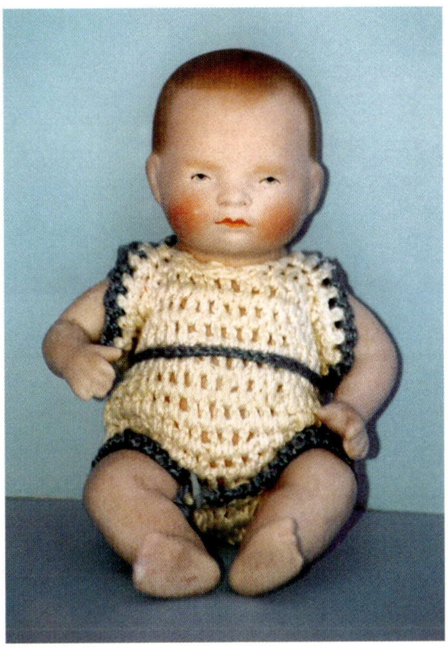

9" tall early all-bisque boy with finely molded hair and features. Jan Foulke in her Kestner book pictures a similar doll that she states could have been a Kestner product. He is not marked. The eyes have a white highlighter mark as well as red eyelid lines over the eye, similar to fine china head dolls. He is jointed at the shoulders and hips and has nicely molded hands and feet as well as his hair and face. (Not enough examples to determine a price). *Courtesy of Paula Reding and the Denver Doll Emporium.*

8.5" tall all-bisque German baby doll with curved bisque arms and legs. She has glass eyes, open-closed mouth, and a mohair wig. The doll is jointed at her shoulders and hips. She is marked "150/2/0" on the back of her head. One of her hands has been repaired. The doll may have been made by the German firm of Hertal, Schwab & Co. ($250+).

5.5" tall all-bisque Bye-Lo Baby, circa mid 1920s. It is marked on the back "20 – 13 1/2/Cop By/G.S. Putnam/Germany." Many of these dolls were made by the German Kestner firm. The all-bisque dolls were made in sizes from 4" to 8" tall. The painted eye version is the most common. This example has painted molded hair and features and is jointed at the shoulders and hips. All-bisque dolls were also produced with wigs and glass eyes, or with glass eyes and swivel necks, or with molded shoes and stockings. This doll has been redressed. ($200+).

7" tall pair of all-bisque (white) baby dolls marked on the back with what appears to be MB in a circle (Morimura Brothers) along with "Made in Japan." The dolls have molded and painted hair and features with open mouths. They have curved bisque infant arms and legs. Both babies are wearing nicely made long baby dresses, probably sewn by a mother for her daughters, when the dolls were new. ($50+ each).

"My Dolly's Nursing Set," with box measuring 7" x 5.25". It includes a 4" tall bisque baby doll. The bottle is marked "Made in Japan." The doll is jointed at the shoulders and hips. The set also includes a white bottle marked on the front "Milk A" and on the back "5 Japan," plus a small clear glass bottle with "Baby Bunting Dolly" marked on one side and a black rubber nipple at the top. A brass colored rattle with a wire handle marked "Japan" and a pacifier with a red rubber nipple are also included. (Set $75-100). *Doll and photograph from the collection of Marilyn Pittman.*

Bisque Head Baby Dolls

Although little girl dolls and even lady fashion dolls with bisque heads had been made since the 1870s and 1880s in France and Germany, baby dolls with bisque heads were not produced until approximately 1910. These baby dolls have now become very collectible, especially those made using unusual designs. Some of the most popular include Bonnie Babe by Georgene Averill, Baby Stewart produced by Gebrüder Heubach, Kämmer & Reinhardt's #100, J.D. Kestner's Hilda and Smiling Century Baby, the Recknagel babies with molded caps, Swaine and Co.'s Lori dolls, Horsman's Tynie Baby , J.L. Kallus' Baby Bo Kaye, Amberg's Vanta Baby, and the various Bye-Lo Babies designed by Grace Storey Putnam. Collectors seem to like the baby dolls with individual "personalities" and therefore the prices rise accordingly on these particular dolls.

The baby dolls from doll makers who are known for quality like J.D. Kestner and Kämmer & Reinhardt are also in demand. Simon & Halbig often made the heads for Kämmer and Reinhardt.

A 1910 advertisement that appeared in *Playthings* magazine pictured the #100 Kämmer & Reinhardt baby doll on a ball-jointed composition body. The *1914 Marshall Field & Co. Kringle Society Doll's Catalog* featured quite a number of bisque head baby dolls. Most of these were 7" tall or smaller and many were in the all-bisque style. In the 9" to 22" sizes, all of the dolls came with moving eyes, mohair wigs, and bent limb composition baby bodies. Instead of being dressed in long baby dresses, all but one of nine models of baby dolls were dressed in short dresses or romper type clothing. The 22" doll came in a long flannel dress. The wholesale price for this doll was $27.00 for a dozen dolls. This cost was less than that for the 15" tall baby doll in a fancier short dress. This model cost $36.00 for a dozen dolls.

During the next few years, German baby dolls were not imported to America because of World War I. The Fulper Pottery Co. tried to help fill the void by producing several different models of dolls including baby dolls. (See "World War I Related Dolls" chapter.) Bisque head dolls from Japan were also imported to fill the void.

By the time German bisque dolls once again appeared in the Montgomery Ward and Sears catalogs, the American made composition dolls had taken over most of the market. Although many bisque little girl dolls were offered for sale, only two or three models of baby dolls with bisque heads, wigs, and sleep eyes were offered for sale in each catalog.

Perhaps the lack of business in the ordinary bisque baby dolls inspired some of the more unusual designs that originated during the 1920s. These included The Bye-Lo Baby, circa 1923, Bonnie Babe in 1926, Century Doll Co.'s baby dolls made by Kestner in 1925, Dream Babies by Armand Marseille, circa 1923-1926, Baby Bo Kaye in 1925, Horsman's Tynie Baby, circa 1924, and Vanta Baby, circa 1927.

The 1927 Kämmer & Reinhardt catalog still featured a line of baby dolls with bisque heads. These included mulatto babies and three different models of newborn baby dolls with painted hair and sleep eyes. It appears that these dolls have cloth bodies and composition hands. The greatest amount of attention is paid to a baby doll with a composition bent leg baby body and a bisque head. It is called Me-Lie-Ba and was the firm's new special baby for 1926. The text states that the doll has a fine Simon & Halbig bisque head, sleeping eyes with eyelashes, movable tongue, a wig of good angora mohair, and is wearing a cambric chemise. The doll looks very much like the #126 Kämmer & Reinhardt baby pictured in this chapter (made by Simon & Halbig). Other baby dolls offered included a baby boy with interesting molded hair on his bisque head and baby dolls with "life-life roguish eyes" (flirting).

Even as late as 1930, some high end stores were still importing bisque head baby dolls from Germany. The FAO Schwarz Christmas catalog from that year offered a baby doll with a bisque head (looks similar to #126) featuring a short mohair wig, sleep eyes, eyelashes, and a composition bent limb baby body. The doll came in sizes from 12" to 21" tall and was priced from $7.50 to $14.50 dressed in a long white organdy dress and a lace cap. The Bye-Lo Baby doll was also still being offered. It had a bisque head, moving eyes, and celluloid hands, and was wearing a long white lawn dress and bonnet. Sizes included 11", 14", 16", and 18", priced from $3.00 to $12.00 each.

The Schwarz catalog also pictured several hampers and sets that featured German bisque head baby dolls. The larger dolls were from 9" to 10.5" tall. The sets sold for $20.00-24.00 each, and came in hampers with trays filled with doll clothing and accessories. The dolls had sleep eyes and mohair wigs.

In 1928, Sears pictured a bisque infant doll with sleep eyes, molded hair (in the "new baby" style), open mouth with two teeth, cotton stuffed body, and a crier. The description states that the doll was Sears' own import, so that was why the price was only 98 cents for the 17.5" tall doll (which may have included its long lawn dress).

The latest quality bisque head baby doll offered by Sears was advertised in 1926. The advertisement stated that the baby had a Kestner head. It was 14" tall with sleep eyes, molded hair, and a cry voice. The doll was dressed in a long white organdy dress and sold for $1.98.

Today, interest in the specially designed German baby dolls holds steady, as do most of the prices. Some collectors may have to "make do" with the large variety of fine quality bisque head babies that have no special names. Every antique doll collector should have a cradle located somewhere in their home that is filled with a variety of bisque head baby dolls. These bisque head baby dolls could make an interesting and unusual collection by themselves.

Louis Amberg & Son

The Louis Amberg & Son firm was located in New York City beginning circa 1907. They are best known for their composition dolls (see "Composition Dolls" chapter), but the company was also responsible for the New Bornbabe baby doll with a bisque head beginning in 1914. This doll looks very much like the Grace Putnam Bye-Lo Baby doll, but it did not secure the attention or the sales of Putnam's baby doll. Amberg's dolls came in sizes from 9" to 17". Amberg also had their 1927 composition head Vanta Baby produced in Germany with a bisque head. The dolls were made in 10" or 20-22" sizes.

9" tall "Newborn Babe" doll, circa mid-teens, made by Armand Marseille for Louis Amberg & Son. The doll has a bisque head with a flange neck, sleep eyes, open mouth with two teeth, and a cloth body with composition lower arms. She is marked on the back of her head "L.A.& S./371. 3/0 D.R.G.M./ Germany." She has been redressed. ($150-200).

Averill Mfg. Co.

The Averill Mfg. Co. began business in New York City circa 1915. The company was well-known for their composition dolls in the early years (see "Composition Dolls" chapter) but they also were responsible for the production of one of the most sought after bisque head baby dolls, the bisque Bonnie Babe. The dolls came in sizes from 12" to 22" tall. An all-bisque Bonnie Babe was also produced in 5" and 7" sizes.

18" tall Bonnie Babe, circa 1926, marked "Copr by/Georgene Averill/Germany/1005/3652/1386." She has a bisque flange head, molded hair, glass eyes, and an open smiling mouth with two lower teeth. The doll has a cloth body with full composition arms and legs to above the knee. The doll was designed by Georgene Averill and, according to Jan Foulke writing in the *16th Blue Book*, the heads were made by the German firm of Alt, Beck, and Gottschalck and the bodies by K & K Toy Co. Bonnie Babe was also produced in an all-bisque style. ($900-1200). *Doll courtesy of Paula Reding and the Denver Doll Emporium.*

Bye-Lo Baby

The bisque head Bye-Lo Baby dolls were made in Germany beginning circa 1923. The dolls were designed by Grace Storey Putnam from the United States. The doll had been modeled by Putnam from a three-day old infant. Many other doll firms began marketing similar dolls when the baby doll proved successful. The Bye-Lo dolls look similar to Amberg's earlier Newborn Babe, first produced in 1914. German firms involved in making the Bye-Lo at one time or another included J.D. Kestner, Alt, Beck, & Gottschalck, Hertel & Schwab, and perhaps others. The cloth bodies and assembly were done by K&K Toy Co. in New York. This company was a subsidiary of George Borgfeldt & Co., the distributor of the dolls. The dolls were still being sold in the early 1930s by FAO Schwarz in sizes of 11", 14", 16", and 18". These dolls were priced from $3.00 to $12.00 each.

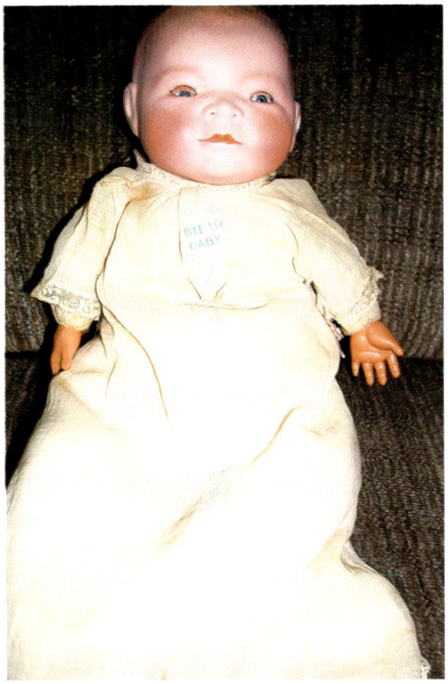

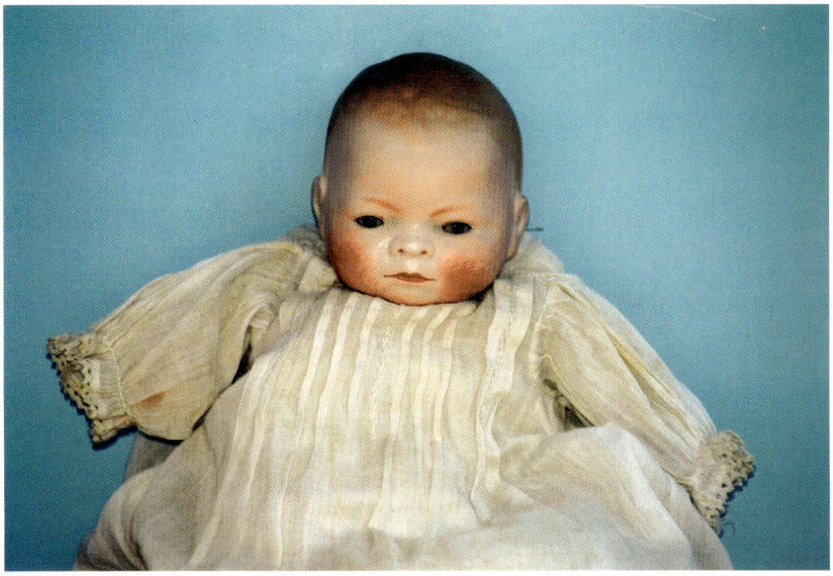

12" tall bisque head Bye-Lo doll showing a close-up of the face. She has molded hair, blue sleep eyes, and her original cloth body with celluloid hands. ($200+).

15" tall Bye-Lo Baby doll with a bisque flange head, circa 1923. She has molded hair, sleep eyes, and a closed mouth. Her body is cloth and is stamped "Bye-Lo Baby/Pat. App. For/Grace Storey Putnam." Her limbs are also cloth and her hands are made of celluloid. She has a working crier in her stomach. She wears her original dress marked "K7K/Reg. U.S./Pat. Off./Bye-Lo/Baby/Not Genuine/Without/Signature/Grace Storey Putnam." The last line is a replica of Putnam's signature. The back of her head is marked "© 1923 by /Grace S. Putnam/Made in Germany." ($300 with original tagged dress and body). *Doll from the collection of Leora Moberly. Photograph by Gail Carey.* (Doll is a replacement for the Bye-lo doll Moberly owned as a child. It was lost on a streetcar in St. Louis when she was young.)

Hertel Schwab & Co.

The German firm of Hertel Schwab & Co. began business around 1910. The company produced a large assortment of baby dolls including all-bisque dolls in 6" and 11" sizes. Another model that is especially desirable is #159, since it has two faces.

19" tall German Hertel, Schwab & Co. baby doll with a bisque socket head, circa 1910. She is marked "Made in Germany/152/11" on the back of her head. The wigged doll has glass sleep eyes and a large open mouth with four teeth. She has a deeply molded chin mark as well as a mold mark to indicate her neck. Her head is mounted on a composition baby body with curved arms and legs. She is jointed at the neck, shoulders, hips, and wrists. She has been redressed ($350+).

Gebrüder Heubach

The German Gebrüder Heubach company began making doll heads around 1910. Most of their dolls were marked with "Heubach" in a square or with the sunburst over a half circle that included the G.H. initials. The firm made several baby dolls. The most popular are their Baby Stuart dolls, which have molded bonnets on their heads. Other Heubach babies include #6894, 7602, 6898, 7759, and 7604. Included are laughing and pouty babies as well as #8420, a wigged baby doll with glass eyes.

8" tall "Baby Stuart" produced by the German Gebrüder Heubach firm, circa 1912-1915. The doll has a molded stationary cap on her bisque socket head. The cap is decorated with painted flowers. Her features are molded and painted, including intaglio eyes and a closed mouth. The bisque head is mounted on a five-piece composition body jointed at the shoulders and hips. She is marked "7977" on the back of her head and has been redressed. The doll was also made with glass eyes and a removable bisque bonnet. ($800+). *Courtesy of Paula Reding and the Denver Doll Emporium.*

Hertwig & Co.

The German Hertwig & Co. firm produced many different designs of dolls during the 1910-1930 years. These dolls included all-bisque, china heads, plus a variety of bonnet head and novelty dolls including pincushion dolls. Very few baby dolls were produced. A bonnet head baby was made circa 1910 and all-bisque babies were part of the company line circa 1920.

8" tall bisque shoulder head doll made by the German firm of Hertwig & Co. She wears a white molded baby bonnet trimmed on each side with a large pink bow. Her features are molded and painted. She has a cloth body and bisque lower arms and legs. The body pictured on the doll in the Hertwig & Co. Archives book by Florence Theriault is cloth with fatter pink bisque lower arms and all cloth legs. Although this body was made commercially, it may be a replacement. She is marked on the back of her shoulder plate "Germany." ($125+).

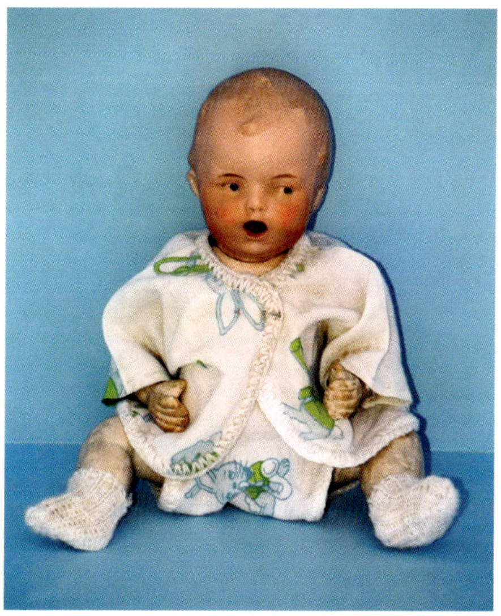

8" tall German Gebrüder Heubach baby with bisque socket head and five-piece papier mâché baby body. He has molded hair and painted features with an open-closed mouth. The back of his head is marked "5/0D/87 Heubach" (in square) "29/Germany." The second set of numbers are hard to read and may not be correct. He has been redressed. ($250+).

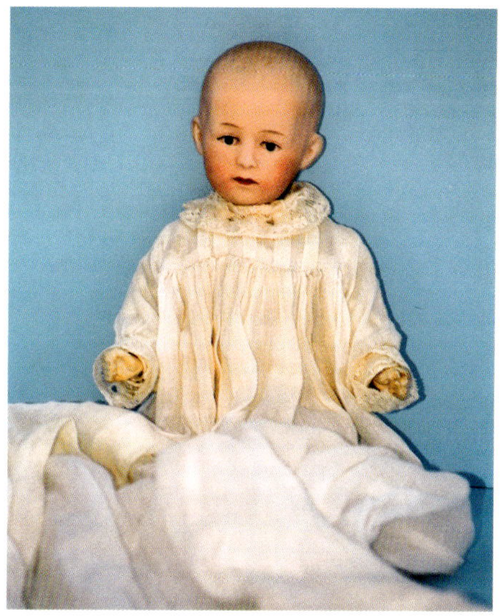

9.5" tall Gebrüder Heubach baby mold # 7602. He is marked with the Sunburst mark plus "7602/ Germany." He has lightly molded hair, painted intaglio eyes, and a closed mouth. This model number head was also used as a little boy. He has been redressed. ($200-250).

Kämmer & Reinhardt

The German firm of Kämmer & Reinhardt began the production of their first baby doll in 1909. It was the popular mold #100, sometimes called the Kaiser baby by collectors. Many of the firm's bisque heads were produced by Simon & Halbig, also located in Germany. Most of the baby dolls issued by Kämmer & Reinhardt were styled with wigs and glass eyes, but some of their babies featured painted eyes and molded hair (#100, 127, 171, 172). The company also produced some unusual mulatto dolls pictured in their 1927 catalog.

Some of the firm's other model numbers for their babies include the following: 115A, 116A, 118A, 119, 121, 122, and 126. Many of these heads were also used on toddler bodies. The dolls are marked "K (star symbol) R" on the back of their heads. The "Simon & Halbig" mark is also sometimes found on the dolls. By the late 1920s, the K & R doll heads were more often made of celluloid rather than bisque.

J.D. Kestner

Of all the dolls made in Germany, those produced by the J.D. Kestner firm are among the most popular with today's collectors. Like other companies, the firm apparently did not begin the production of baby dolls until 1910. Jan Foulke, in her book *Kestner, King of Dollmakers*, describes the early Kestner advertising involving baby dolls. She states that what she believes is a mold #211 baby doll was advertised in 1912 as "Natural Baby." By January of 1913, a *Playthings* ad described "Kestner character babies [as] the great hit of the past holidays."

The most sought after baby doll appears to be Kestner's "Hilda." She was made in both infant and toddler designs. These dolls are marked with both the Hilda name and mold #245.

Kestner also produced all-bisque baby dolls as well as several models of shoulder head character babies (209, 210, 234, 235). Other Kestner mold numbers for some of their baby dolls, identified by Foulke, include: #220, 226, 236, 247, and 262. All of these dolls have glass eyes and wigs. Other collectible Kestner babies include a Kestner oriental baby, mold number 243, and the O.I.C. screaming baby, which has also been attributed to Kestner. Another special baby was marked only J.D.K. Collectors sometimes refer to this doll as Baby Jean. She has painted hair. The "bald" head babies have no mold numbers, only size marks appear on the back of their heads. These dolls came mostly with glass eyes but a few had painted eyes. The Sears catalog was still offering a baby doll with a Kestner head as late as 1926. Kestner also made bisque baby doll heads for the New York-based Century Doll Company beginning in 1925. The babies came with glass eyes and either closed mouths or an open-closed mouth with molded teeth.

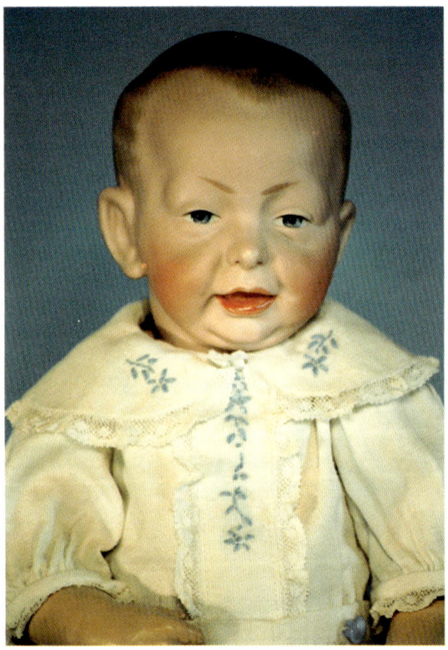

15" tall German Kämmer & Reinhardt #100 baby doll, sometimes referred to by collectors as the Kaiser baby. He has a bisque socket head with molded painted hair, painted features, and an open-closed mouth. His head is mounted on a five-piece composition bent limb baby body. He is marked on the back of his head "36/K (star symbol) R/100." He has been nicely redressed. ($450-500).

16" tall baby doll produced by the German J.D. Kestner firm. According to Jan Foulke writing in her book, *Kestner; King of Dollmakers*, this was the first Kestner head to carry the "J.D.K." initials, which stood for Johannes Daniel Kestner. The doll is marked "J.D.K. 211." She has a bisque socket head, which is mounted on a five-piece composition baby body. The doll has a mohair wig, glass eyes, and an open mouth with bottom teeth. She has been nicely redressed ($600+). *Courtesy of Paula Reding and the Denver Doll Emporium.*

18" tall Kestner baby doll, mold number 226. She has a bisque socket head mounted on an excellent five-piece composition baby body. The baby has a mohair wig, glass sleep eyes, and an open mouth with two teeth in the top of her mouth. She has been redressed. ($700). *Denver Doll Emporium.*

13" Kestner character "Hilda" baby doll with a bisque socket head, circa 1914. This doll is one of the most sought after of the baby dolls. She has a mohair wig, glass brown eyes, and an open mouth with two teeth. The head is mounted on a five-piece bent limb composition baby body. Hilda is marked "Made in/Germany/245/J.D.K./1914 /©/Hilda." The doll also came in a black model. This example is very nicely dressed. ($2,200+). *Denver Doll Emporium.*

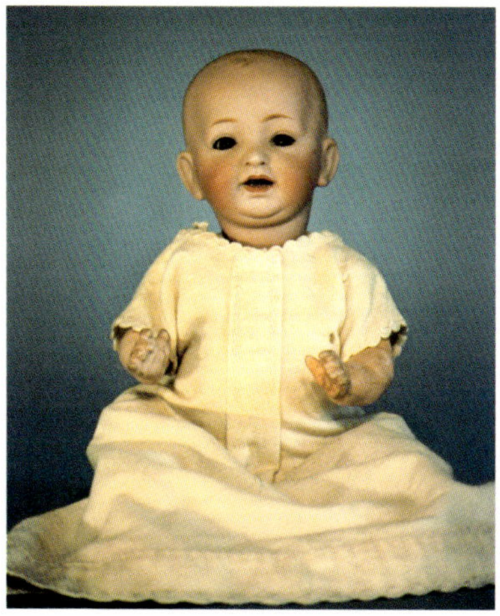

11" tall Kestner baby doll with a bisque socket head and molded hair. He is marked "J.D.K./7." The 7 is the size number of the doll. The molded hair Kestner baby dolls did not have mold numbers. The head is mounted on a five-piece bent limb baby body. He has brown stationary glass eyes, an open mouth, and two molded teeth. The same model was also made with painted eyes. He has been nicely redressed. ($300-350).

16" tall "Smiling Century Baby" made by Kestner for the New York-based Century Doll company beginning in 1925. Most of the dolls had closed mouths and were designed to look as much like the popular Bye-Lo Baby as possible. This baby has an unusual smiling face. The bisque head is mounted on a cloth body with composition hands. It is marked "Century Doll Co./Kestner Germany." The doll has molded hair, glass sleep eyes, and an open-closed smiling mouth with two molded teeth. ($650-750). *Courtesy of Pauling Reding and the Denver Doll Emporium.*

15" tall unusual character baby doll, attributed to Kestner, with a bisque flange head. The doll is unusual because his open-closed mouth is molded in a crying-screaming position. No doubt little girls would have quickly tired of playing with this baby doll. He has molded hair and tiny glass eyes about the size of pinheads. The doll has an all-cloth body except for composition hands. He is marked "255/2/O.I.C./Made in Germany." ($1,200). *Denver Doll Emporium.*

Armand Marseille

The German Armand Marseille firm must have been the most prolific doll maker in Germany. Besides producing their own doll heads, they sometimes provided heads for other companies as well. Their dolls ranged in quality and design from average to excellent. Most of their baby dolls were so popular at the time they were made that today's collectors can find examples at reasonable prices. The firm produced dolls with both molded hair and fitted wigs.

Some of the firm's named baby dolls include Kiddiejoy, Dream Baby, Gloria, and Phyllis. Other bald head mold numbers include #341, 351, and 352. Mold numbers for baby dolls with wigs are #326, 590, 985, 971, 996, 990, and 991.

Dream Baby is sometimes found marked Arranbee because that American firm marketed this doll in their early years of business.

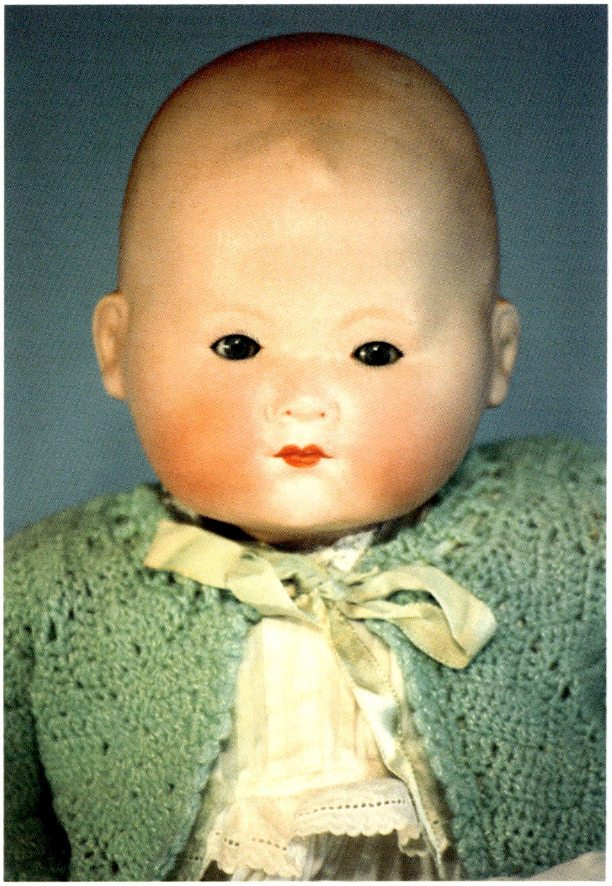

20" tall Armand Marseille Kiddiejoy baby doll. Visible marking reads "Germany/ Kiddiejoy." She has a bisque flange head, molded painted hair, sleep eyes, and a painted closed mouth. The head is mounted on a cloth body with swing hip joints, cloth arms, and hands of composition. She has been nicely redressed. The Jacobs & Kassler firm of New York City sold the dolls. They were featured in an ad in *Playthings* in May 1927. ($350+).

8" and 11.5" tall "My Dream Babies" made by the German Armand Marseille firm, circa 1926. The same doll head was also used for the American Arranbee doll from circa 1926. This is when the dolls were given the My Dream Baby name. Most of these dolls were made with bisque flange heads, inexpensive cloth bodies, and composition hands. The dolls have molded painted hair, sleep eyes, and painted closed mouths. The smaller doll is equipped with a squeaker that no longer works. The marks on the smaller doll read "A.M./Germany." A mold number may be hidden by the body. The larger doll is marked "A.M./Germany/341/O.K." Both dolls have been redressed. (8" $75-100, 11" $125+).

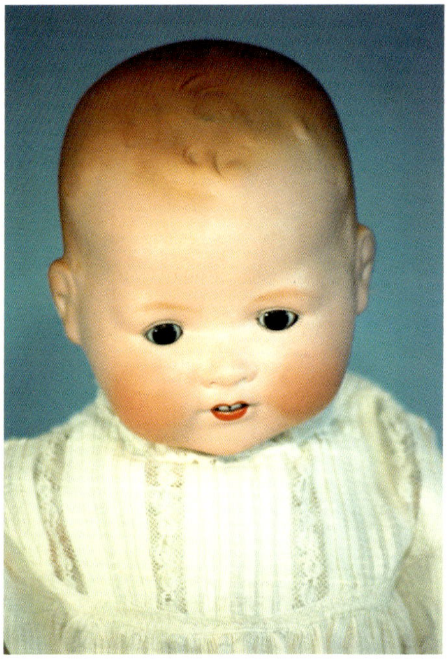

17" tall baby doll made by Armand Marseille, mold number 351. She has a bisque socket head with molded hair, brown glass sleep eyes, and an open mouth with two teeth. The head is mounted on a five-piece composition bent limb baby body. She is marked "A.M./Germany/351/5.K. ($350+).

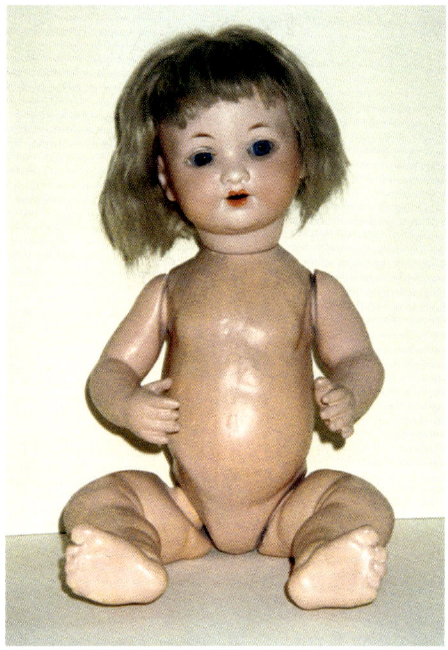

16" tall Armand Marseille baby doll, #985. She is pictured without her very long baby dress so her face and body can be seen. She has a bisque socket head, blue sleep eyes, an open mouth with two upper teeth, and a mohair wig. Her five-piece bent limb baby body completes the doll. She is marked on the back of her head "A985M/Germany/4 ½." ($250+). *Doll and photograph from the collection of Marilyn Pittman.*

9" tall Armand Marseilles Googly doll with bisque socket head and a nice five-piece composition baby body, circa 1920. He has molded hair in a top knot, molded and painted features with a closed mouth and eyes looking to the side. He is marked on the back "GB Germany/DRMR" (perhaps for George Borgfeld, distributor). The doll is also marked "A252M." ($1,200-1,400). *Denver Doll Emporium.*

10" tall Recknagel doll with a bisque socket head like the previous example except the bonnet is trimmed in pink molded bows. This doll has a five-piece toddler body. She is marked "RA 202 7-0." The doll has been redressed. ($450-500). *Denver Doll Emporium.*

Recknagel

The German Recknagel firm produced a number of small bisque dolls around the period of World War I. Most collectible are their line of bonnet babies. These range in size from 6" to 10" tall. Most of the bodies for these dolls are rather crude, having been made of a papier mâché type of material. Some of the dolls have bare feet and others feature painted shoes and socks.

The firm produced a number of small baby dolls without bonnets as well. Most of the company's dolls are marked "RA" with numbers following.

8" tall Recknagel baby doll with a bisque socket head wearing an example of a molded bisque bonnet. It is trimmed with three blue molded bows. The doll has molded hair, intaglio eyes, and an open-closed mouth with two molded teeth. She has a five-piece composition baby body. She is marked "22/1 2/0" and has been nicely redressed. ($375-400). *Courtesy of Paula Reding and the Denver Doll Emporium.*

8" tall baby doll with bisque socket head made by the German Recknagel firm. His white molded cap has a pink molded top knot for trim. He has molded hair, painted eyes, and an open-closed mouth with two molded teeth. The doll has an inexpensive five-piece papier mâché type composition body. The back of the head is marked "RA/28-12." He wears his original baby dress. ($275).

6.5" tall Recknagel toddler doll wearing a molded blue ruffled bonnet. Her bisque socket head has molded hair under her bonnet, eyes looking to the side (once known as peek-a-boo eyes), and a painted closed mouth. Her body is made of an inexpensive type of papier mâché with molded shoes and socks. She is marked on the back of her head "R 39 A." She has been nicely redressed. ($450+). *Denver Doll Emporium.*

Simon & Halbig

The German Simon & Halbig firm produced fine bisque doll heads to be used on its own dolls as well as on dolls marketed by other firms including Kämmer & Reinhardt. The K & R company purchased the Simon & Halbig firm in 1920.

Some of the wigged babies included Erika #1489 (with wig and glass eyes), #1294, 1428, and 1488. This firm also made bald head babies like mold #1498, but most of their babies had fitted wigs. Both types of babies had glass eyes.

15" tall bisque socket head baby doll produced by Simon & Halbig for the Kämmer & Reinhardt German company. She has sleep eyes with real eyelashes, an open mouth with two upper teeth, and her original mohair wig. The head is mounted on a five-piece bent limb composition baby body. She is marked on the back of the head "Simon & Halbig/126/36." The Kämmer & Reinhardt marking appears to be under her hair and not visible. She has been nicely redressed. ($400+).

Swaine & Co.

The bisque head character babies made by the German Swaine & Co. have become very popular with today's collectors. The Lori dolls are especially desirable. They include an incised Lori with molded hair, glass eyes, and an open-closed mouth, as well as an open mouth Lori marked #232.

Other Swaine baby dolls include one marked DI that features molded hair, an open mouth and intaglio eyes, plus a character baby with a wig, glass eyes, and closed mouth marked DIP, as well as the pictured DV baby.

The Swaine mark includes the words "Geschutzt/S & Co./Germany" printed inside a circle.

16" and 13" tall bisque socket head baby dolls, both made by the German firm of Swaine & Co. The larger doll has molded hair, glass eyes, and an open-closed mouth. She is marked "DV/6." She has also been stamped "Geschutzt/S & Co./Germany." The smaller doll is an open mouth Lori. She has molded hair, glass eyes, and an open mouth with bottom teeth. She is marked "232/2." Both dolls have five-piece bent limb baby bodies and both have been redressed. (13" $800+, 16" $1,200+). *Courtesy of Paula Reding and the Denver Doll Emporium.*

Japanese Dolls

During World War I, several Japanese manufacturers were called on to produce bisque head dolls to replace the ones no longer available from Germany. Their first efforts were not too successful, but the later dolls attained similar quality to those made, for the same reason, by the Fulper Pottery Co. in the United States. Some of the best of the baby dolls were imported by New York distributors, especially Morimura Brothers. The dolls this firm handled were marked MB in a circle along with Japan. Other dolls were marked "Nippon" instead of Japan.

14" tall bisque socket head baby doll made in Japan. He is marked with MB in a circle mark (Morimura Brothers) plus "2 Japan 5." He has molded hair, glass eyes, and an open mouth with teeth. The head is mounted on a five-piece composition bent limb baby body. ($150).

12" tall bisque socket head baby doll marked "Nippon #503" He has molded hair, glass eyes, and an open mouth with molded teeth. The doll has a five-piece bent limb composition baby body and has been redressed. ($125-150).

175" tall Japanese bisque socket head baby doll on a composition baby body with curved arms and legs. The hands are attached separately. She has stationary glass eyes and an open mouth with two teeth. The wigged baby is marked on the back in a circle MB along with a straight vertical line with a V shape at the top and bottom of the line. "Japan/2/0" completes the marking. She was also sold by Morimura Brothers. The doll originally belonged to a lady born in 1908, so the doll is circa World War I years. The author was named after this young woman, who died at a young age. (Doll $125, folding bed $35).

Bisque Head Character Dolls

Some of the bisque dolls most popular with current collectors are those called character dolls. Many of these types of dolls were produced in Germany (and sometimes France), beginning in 1910. They have bisque heads, wigs or molded hair, and painted or glass eyes. Their bodies can range from the German composition ball-jointed styles to less expensive five-piece composition or papier mâché examples. The dolls were not made with the usual "dolly" faces. Instead, each doll model had an individual look. Sometimes the difference might have been created by including an unusual hairstyle (a wig with braids over each ear or a molded hair ribbon), big googly or flirting eyes, or an expression such as a pouty or smiling look. Dolls that could "do things" were also considered character dolls. For example, Gebrüder Heubach made a boy doll that could whistle and another that clapped his hands. Unusual dolls were also produced by several companies that manufactured "walking" dolls.

Some of the German firms that produced bisque head character dolls during this period included Alt, Beck & Gottschalck, Catterfelder Puppenfabrik, Goebel, Hertel, Schwab & Co., Ernst Heubach, Gebrüder Heubach, Kämmer & Reinhardt, J.D. Kestner, Kley & Hahn, Armand Marseille, Gebrüder Ohlhaver, Recknagel, Bruno Schmidt, Simon & Halbig, and Adolf Wislizenus. The French S.F.B.J. concern also marketed these types of dolls. The character dolls probably did not stay in fashion very long because of World War I. The war affected the German doll industry for several years after it began in 1914, and it was not until the early 1920s that German exports to the United States began to grow. By that time, the composition doll companies located in the U.S. had gained a foothold in the doll market and these local companies continued to prosper. In an effort to regain business, the German doll firms began making "special" dolls that had ties to the United States. These included the Kewpies, Bye-lo Babies, Bonnie Babe, Mibs, Baby Bo Kaye, Baby Peggy, and Tynie Baby as well as others.

Ernst Heubach

Although many of the dolls produced by the German Ernst Heubach firm are of lesser quality, some of their character dolls are very collectible. These include model #276, which had a bug on the nose, plus dolls with molded hair and painted eyes, including models #261, #262 (pictured here), #271, and the toddler with glass eyes and a wig #320. The firm's products were marked in a variety of ways, including a horseshoe surrounding the letters "EK" and with the names "Heubach-Kopplesdorf."

7.5" tall boy doll with peek-a-boo eyes. These types of dolls were advertised in 1914. He has a bisque socket head mounted on a five-piece composition body jointed at the shoulders and hips. The doll has molded hair and painted eyes, and is biting his lower lip. His shoes and socks are molded and painted. His markings read "Germany/E.H. 262-14/0/DRGM." He was made by the German Ernst Heubach firm. ($350-400+). *Courtesy of the Denver Doll Emporium.*

Gebrüder Heubach

The German Gebrüder Heubach firm began producing doll heads around 1910. This time period coincided with the beginning of production for both the bisque baby dolls and the newly styled character dolls featuring unusual bisque heads. Collectors of character dolls have many different Heubach models to choose from. The firm produced both girls and boys in their wide variety of designs. Most of these dolls included molded hair and intaglio eyes. A lesser number were made with wigs and glass eyes. The firm issued dolls with either bisque shoulder heads or socket heads. Sometimes the Heubach bodies were very crude, while other Heubach dolls have heads mounted on nice composition baby bodies or German ball-jointed bodies. Over fifty mold numbers belonging to the Heubach firm have been identified. The dolls are marked either with "HEUBACH" in a square or with a sunburst design and the initials "GH."

The *1914 Marshall Field & Co. Doll Catalog* pictures four bisque Heubach character heads that could be purchased without bodies. The two girls shown are Coquette and the head called "Singing Girl" by collectors. The heads were 3.75", 4.75", and 5.5" tall. Only a wholesale price was given, which ranged from $3.00 to $7.20 a dozen. Two boy designs were also pictured. Both of them had molded hair. One appeared to be one of the pouty models and the other had an open mouth with teeth (may have been molded). These heads ranged in size from approximately 3.5" to 5.25" high and were priced from $2.30 to $5.50 per dozen. The fact that these heads were sold separately means that some of the Heubach heads were purchased by consumers to replace broken heads on earlier dolls or to use with bodies produced by a variety of different companies. That could help to explain the variety of kid and cloth bodies found with the shoulder head dolls. Since the heads and kid bodies were also sold wholesale, dolls could have been assembled by a variety of firms and sold as complete dolls. This same catalog also offered a variety of different bodies at wholesale prices. The 1914 catalog also pictures two completed bisque shoulder head character boy dolls that appear to be Heubach products. One has a kid body with "knee and seat incisions" and the other is shoulder, elbow, and hip jointed. The dolls range in size from 12.5" to 14" tall. The lower priced body dolls were sold for $4.20 to $8.00 per dozen wholesale while the jointed body doll sold for $15.00 per dozen. All of these dolls were advertised as having "character faces." Currently, the Heubach dolls with shoulder heads are generally less desirable to collectors than those with socket heads.

11" tall #7850 Coquette by Gebrüder Heubach. She is marked "Heubach" in a square, plus "Germany." She has a bisque shoulder head featuring light brown molded hair with a blue molded ribbon and hair bow, intaglio eyes, and an open mouth with teeth. Her head is mounted on a kid body with lower bisque arms. She has been nicely redressed ($500+).

14.5" tall Gebrüder Heubach's #7764 "Singing Girl." She is marked "Heubach" in a square. She has a bisque socket head plus molded and painted hair accented with a large pink molded hair bow. She has painted features with side-glancing intaglio eyes, and an open-closed mouth. The head is mounted on a composition and wood jointed body with high knee joints and straight wrists that have been repainted. The doll has been nicely redressed ($5,000-6,000). *Courtesy of Frasher's Doll Auctions, Inc., Oak Grove, Missouri.*

13" tall #10557 little girl doll by Gebrüder Heubach. She is marked "Heubach" in a square. She has a bisque socket head featuring a mohair wig, glass eyes, and an open mouth with teeth. Her head is mounted on a five-piece composition body jointed at the shoulders and hips. She has been redressed ($700-900). *Courtesy of Paula Reding and the Denver Doll Emporium.*

11" tall #10582 little girl doll by Gebrüder Heubach. Her bisque socket head is marked "Heubach" in a square as well as "Germany" on the back of her head. She wears a real hair wig, has glass eyes, and a small open mouth with teeth. Her five-piece composition body is jointed at the shoulders and hips. She has been redressed. ($900-1,100). *Denver Doll Emporium.*

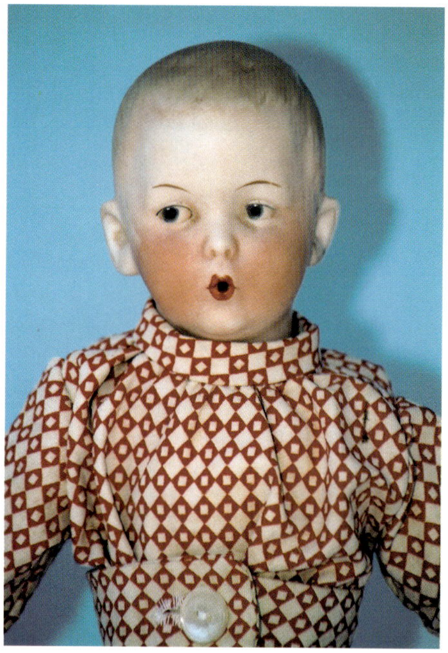

14" tall #8774 whistling doll, sometimes called "Whistling Jim." He was made by Gebrüder Heubach. The back of his socket head is marked "Heubach" in a square. He has molded painted hair, intaglio eyes, and an open mouth molded in a whistling position. He has a cloth body with composition lower limbs. The bellows in his torso made a whistling sound when pressed (not working). He has been redressed. ($700-1,000).

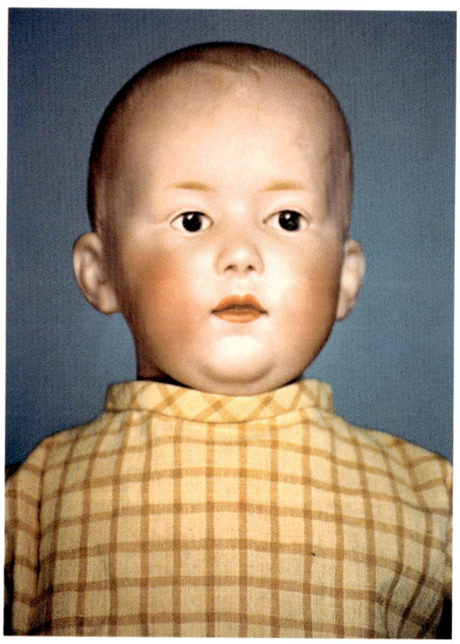

20" tall bisque shoulder head boy doll produced by Gebrüder Heubach. Marked on the back of the shoulder plate is "7/Germany" along with the GH Sunburst mark. He seems similar to the pouty boy #7603. The doll has lightly molded hair, large intaglio eyes with red dots at the inside corner, and a slightly open-closed mouth. The head is mounted on a replaced kid and cloth body with lower bisque hands. The doll has been redressed (with replaced body $350+).

14" tall bisque shoulder head boy by Gebrüder Heubach. He appears to be #7604 but the marking is hard to read. He has lightly molded hair, small intaglio eyes, and an open-closed mouth. The shoulder head is mounted on a kid body with bisque lower arms. ($350+).

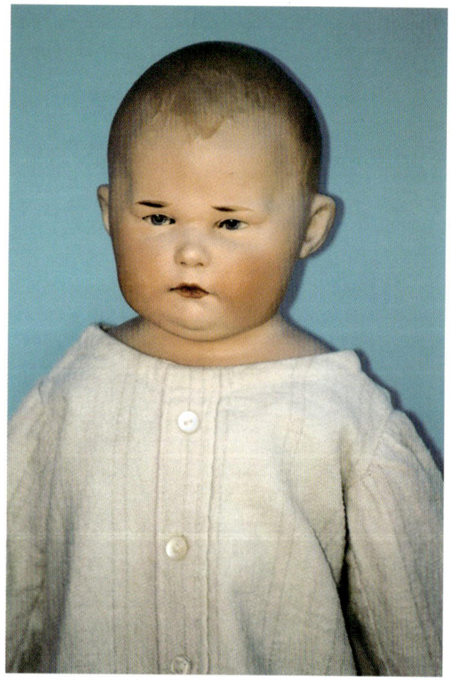

18" tall bisque shoulder head boy doll made by Gebrüder Heubach. He is marked "6/Germany 7846." The numbers are hard to read and the second and last numbers may not be correct. The head is also marked with "Heubach" in a square. He has lightly molded hair, small intaglio eyes, and a slightly open-closed mouth. The doll has a kid and cloth body and lower bisque hands. He has been redressed in a flannel nightgown of the period, as he spends all his time in a circa 1914 cradle advertised in the Marshall Field catalog of that date. (See "Furniture" chapter.) ($500+).

13.5" tall laughing shoulder head boy made by Gebrüder Heubach. He is marked "Heubach" in a square and "Germany." He looks similar to mold number 8191 but he has no lower teeth. The doll has molded brown hair, small intaglio eyes, and a wide open-closed mouth with molded upper teeth. His body is kid and cloth and he has lower bisque arms. The doll has been nicely redressed. ($350).

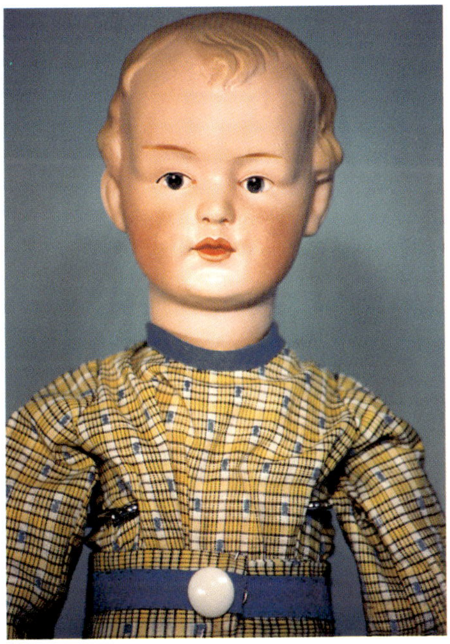

15" tall bisque socket head boy doll #7622. Besides his mold number, he is marked "5", along with the G. Heubach Sunburst symbol, and "Germany." The doll has nicely molded light brown hair, intaglio eyes with red dots in the corner, red dots in his nostrils, and an open-closed mouth. The turning socket head is mounted on a jointed German body with joints at the shoulders, elbows, hips, and knees. The upper legs are long narrow pieces of wood that make the legs look out of proportion to the heavier body. The body is original but cheaply made. He has been redressed. ($700-800).

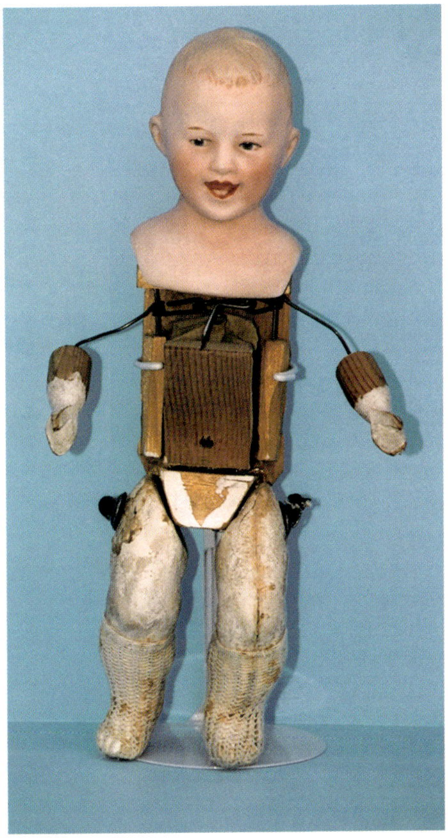

10" tall unusual shoulder head doll made by G. Heubach. He is marked with the Heubach Sunburst mark as well as "Germany" on the back of his shoulder plate. He has light brown slightly molded hair, intaglio eyes, and an open-closed mouth with two lower teeth. He has a crude baby body with a simple mechanism to allow his hands and arms to move when the center piece of wood is pushed. The hands are wood and the legs appear to be made of a papier mâché type material. When he is properly dressed, his unusual body does not show. He wears his original factory-made stockings. ($350-400).

Helen W. Jensen

Helen W. Jensen designed the "Gladdie" doll and it was produced in Germany circa 1928-1930. The heads were made of bisque or biscaloid. The biscaloid material, according to the *Eighth Edition of Doll Values* by Jan Foulke, was an imitation bisque that is like a terra-cotta ceramic. Although the bisque heads were made in Germany, the bodies were produced by K & K for the New York based George Borgfeldt firm that marketed the dolls.

18" tall "Gladdie" designed by Helen W. Jensen with the head made in Germany, circa 1928-1930. The bodies were made by K & K. She has a flange head made of painted biscaloid, molded hair, glass eyes, and an open-closed mouth with teeth. The body is cloth with composition arms and lower legs. The back is marked "Gladdie/Copyriht By/Helen W. Jensen/Germany." Copyright is misspelled. ($800+). *Courtesy of Paula Reding and the Denver Doll Emporium.*

Kämmer & Reinhardt

The German firm of Kämmer & Reinhardt marketed many of the most sought after bisque head character dolls ever made. Many of these bisque heads were actually produced by other firms, including Simon & Halbig. K & R offered their first character dolls as early as 1909. A *Playthings* ad from 1910 (shown in the Colemans' book *The Collector's Encyclopedia of Dolls*) pictured several of the new character dolls. Included were Peter & Marie #101, Elsie #109, Carl #107, Gretchen & Hans #114, and Elsa & Walter. The baby model #100 was also included in the advertisement. All of the dolls had painted features and wigs. Doll model #101 also was made with glass eyes. Peter & Marie came in sizes ranging from 8" to 20" tall, Hans & Gretchen were sized from 8" to 25" tall. This #114 model was also produced with glass eyes. Elise (#109) was made in sizes from 7" to 21" high and #107 Carl came in 12" and 22" sizes.

Some of the other K & R characters include #112 (boy or girl with painted eyes and wig), #115A (wigged toddler with glass eyes), and #102 (boy with molded and painted hair and eyes). The dolls are usually marked "K & R," with the "and" sign generally enclosed in a star symbol. Sometimes the doll's head also includes the letters S & H for Simon & Halbig if that firm produced the head.

8" tall Gretchen & Hans (each mold #114) made by the German firm of Kämmer & Reinhardt. The wigged girl and boy dolls have bisque socket heads with painted eyes and closed mouths. Hans has a fully jointed ball-jointed body while Gretchen's arms do not have the elbow or hand joints. Both dolls appear to be wearing their original underwear. The Gretchen & Hans dolls were advertised in *Playthings* magazine in 1910. The dolls were made in a variety of sizes up to 25" tall. ($1,800-2,000 each). *Photograph and dolls from the collection of Jean M. Grout.*

16" tall bisque socket head character doll by Kämmer & Reinhardt marked "K (star symbol) R/Simon & Halbig 116A." She has brown sleep eyes, open-closed mouth and upper teeth, dimples, and a blonde mohair wig. The head is mounted on a composition ball-jointed toddler body. She has been redressed in a cotton playsuit. (Sold at auction in 2004 for $1,900). *Courtesy of Frasher's Doll Auctions, Inc., Oak Grove, Missouri.*

15.5" tall mold #101 Marie by Kämmer & Reinhardt. She has a bisque socket head with a braided wig, intaglio painted eyes, one stroke eyebrows, and a closed mouth. Marie has a composition ball-jointed body. A Marie doll and the male Peter version of #101 were pictured in a 1910 *Playthings* ad. These dolls came in sizes from approximately 8" to 20" tall. ($2,700 -3,300). *Courtesy of Paula Reding and the Denver Doll Emporium.*

J.D. Kestner

The character child dolls produced by the German J.D. Kestner firm represent the fine designs and workmanship always offered by the "King of Dollmakers" company. In her book, *Kestner, King of Dollmakers*, author Jan Foulke pictures over twenty different models of Kestner character dolls. Even more examples can be added when additional numbers are included to reflect both painted and glass eyed dolls. These dolls can be identified by their mold numbers. The later characters from the 200 series are also marked "J.D.K." along with the mold number.

Like several other German firms, Kestner began production of these dolls circa 1910. They were supposed to have been designed to resemble real children to replace the usual "Dolly Face" dolls. Although these new dolls probably appealed to many parents, little girls no doubt still preferred the "Dolly Faced" dolls that featured sleep eyes and wigs. Schoenhut had to discontinue many of the designs from their first issue of dolls because little girls preferred "pretty" dolls instead of realistic models of children as their playmates.

Only one rare Kestner model of a character doll was made with molded hair. He is #177 and Foulke pictures him in her book. The other Kestner character dolls came with wigs and most had painted eyes. There were some exceptions, as Foulke pictures models #179, 184, 185, 189, 206, 208, and 212 with glass eyes. Mold numbers for the Kestner character dolls range from 177 to 190. Later dolls include the numbers 206, 208, and 212. Kestner character dolls included both boy and girl models and most range in size from 12" to 18" tall.

Armand Marseille

The German Armand Marseille firm was probably the most prolific doll company in Germany during the teens and 1920s. The company produced every type of bisque doll including all-bisque, "Dolly Faced" dolls in both socket and shoulder head styles, character babies, and many different examples of character children. In addition, the company also manufactured bisque heads that were used by other firms. The Armand Marseille dolls are marked with the letters "A.M." somewhere on the back of the doll's head. Some of the dolls also carry their own names, like "Fany." Numbers may occupy the space between the A and the M. Some of the firm's most famous character children include Fany (#230 with molded hair, #231 with wig), and the "Just Me" wigged character dolls with eyes to the side (#310). The firm also produced googly-eyed dolls (#323, 253, 210) and flapper dolls (#400) during this period.

In addition, the firm designed boy dolls with molded hair and eyes, (#500, 560, 600), little girls with closed mouths, wigs, and either glass eyes or painted eyes (#550, #700) and a little boy with a wig and glass eyes (#590).

11" tall Googly-eyed doll #323 made by the German firm of Armand Marseille. Similar dolls were advertised in the Marshall Field doll catalog in 1914. This example has a wigged bisque socket head, large googly eyes and a closed mouth. The head is mounted on a five-piece composition body. The back of the head is marked "Germany/323/A3-0M." She has been nicely redressed. ($1,200-1,500). *Courtesy of the Denver Doll Emporium.*

12" tall mold #185 character girl made by the J.D. Kestner firm. She has a bisque socket head with a mohair wig, painted eyes, and an open-closed mouth with teeth. The doll has a composition ball-jointed body and is wearing her original underwear and hair bows. This model came in sizes from 12" to 18" tall and was also made with glass eyes. ($2,000-2,500). *Photograph and doll from the collection of Jean M. Grout.*

9" tall Googly-eyed doll #253 made by Armand Marseille. She has a wigged bisque socket head, large glass googly eyes, a watermelon closed mouth, and a five-piece composition body. She is marked "253/GB/Germany A6-0M." She has been redressed. ($1,800-2,000). *Courtesy of Denver Doll Emporium.*

16" tall flapper type doll made by Armand Marseille circa mid 1920s. She is marked "Armand Marseille/Germany/400/A01/2M." She has a bisque socket head with a short real hair wig, glass eyes, closed mouth, and a newer styled composition ball-jointed body with the leg joints above the knee. This allowed the dolls to wear shorter dresses to follow the newer styles. ($2,250-2,500). *Courtesy of the Denver Doll Emporium.*

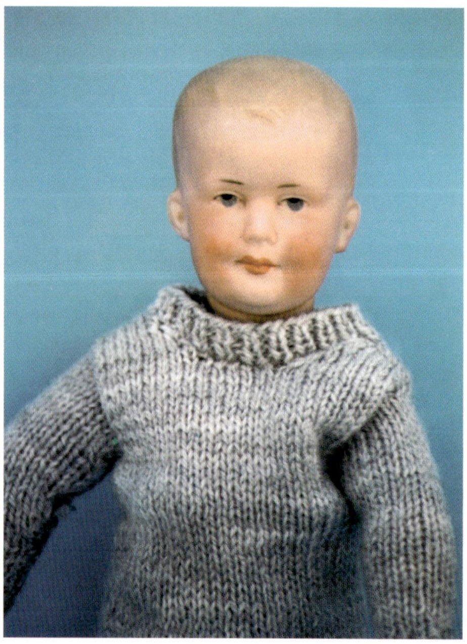

12" bisque character boy, mold #50? (maybe 500), made by Armand Marseille. He has a bisque socket head, molded hair, intaglio painted eyes, a closed mouth, and a composition ball-jointed body. In addition to the mold number, he is marked "Germany/A0M." The doll has been redressed ($400-450).

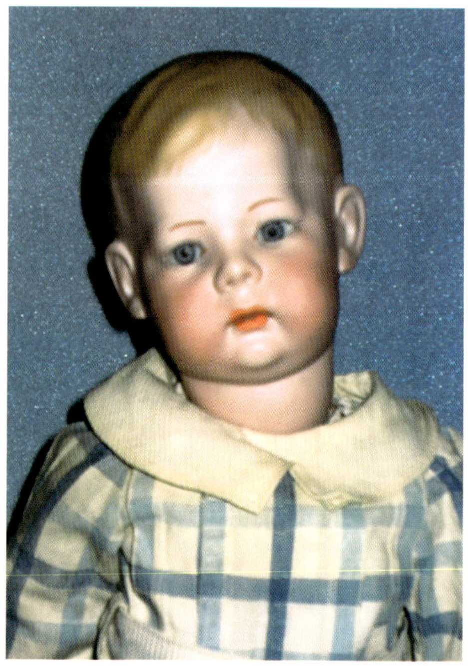

16" tall mold #230 Fany made by Armand Marseille, circa 1912. He has a bisque socket head with molded hair, glass eyes, very slightly open-closed mouth, and a composition ball-jointed body (except at wrists). He has been redressed. ($6,000+). *Photograph and doll from the collection of Jean M.Grout.*

Simon & Halbig

The German Simon & Halbig firm was also a quality maker of bisque dolls, until the firm was purchased by Kämmer & Reinhardt in 1920. The company had already been producing doll heads for K & R, as well as for other companies. Their dolls are usually marked with an "S H" somewhere on the back of the doll's head. Numbers can often be found between the two letters.

Character child dolls from Simon & Halbig included both boy and girl models with painted eyes. The wigged girl pictured here is #151. A similar #150 was also produced. A little girl character toddler doll #1428 with a wig and glass eyes is also most unusual. Boy dolls include #153 with molded hair and painted eyes, and #150 with painted eyes and a wig. A pair of 14" tall matching Simon and Halbig dolls is pictured in Patricia Smith's *Doll Values 9th Edition* book. The dolls are marked "151 S&H 5 ½". They have wigs, painted eyes, and open-closed mouths with molded teeth. One is dressed as a girl and the other as a boy.

8" and 7.5" tall "Just Me" #310 dolls made by Armand Marseille beginning circa 1929. The doll on the left has a bisque socket head on a five-piece composition body. She is wigged, has glass sleep eyes, and a closed mouth. She is marked on the head "Just Me/Registered/Germany/310/10/0." She is dressed in a tulle and silk dress by Vogue Dolls, Inc. and wears original leather shoes with pom poms circa 1930s. The doll on the right has a painted bisque socket head with a composition body, glass sleep eyes, closed mouth, and a mohair wig. Her head marking is the same except for "11/0." She has been redressed in a reproduction of one of Vogue's original designs. (Bisque $1,400-1,600, painted bisque $750-850). *Photograph and dolls from the collection of Carol Stover.*

16" tall bisque socket head character doll #151 made by the German Simon & Halbig firm. She has a mohair wig, painted side glancing eyes, open-closed mouth, and molded teeth. The head is mounted on a composition ball-jointed body. She is marked "151 S&H 0 ½." ($4,000-5,000). *Courtesy of Frasher's Doll Auctions, Inc., Oak Grove, Missouri.*

S.F.B.J. (Socíeté Francaise de Fabrication de Bébé's & Jouets)

The S.F.B.J. organization was formed in Paris, France, circa 1899. It was made up of several different French doll makers, including Jumeau and Bru. An assortment of character dolls were marketed by this group. The organization marked their dolls, "S.F.B.J." along with a model number and "Paris 4." Their dolls were produced in both wigged and molded hair styles. Either painted or glass eyes were used on the dolls. Most of the dolls had glass eyes and wigs.

Boy dolls with molded hair and painted eyes include model #226, 233 (crying), and 235. Boys with molded hair and glass eyes include #227 and #238. A boy doll with a wig and glass eyes is model number 236. Flocked hair was used on the boy doll carrying the number 237.

Girl dolls were produced with glass eyes and wigs in model numbers of 236, 238, 248, 251, and 252. Model number 245 was a googly doll with large glass eyes and a wig.

S.F.B.J. marketed a large number of character faced dolls in a variety of styles. These dolls ranged in size from 12" to 22" tall.

12" tall bisque socket head little boy doll marked "S.F.B.J./238/Paris 4/" on the back of his head. He has lightly molded and painted hair, jewel eyes, an open-closed mouth with molded teeth, and one stroke brows. His body is composition. He has been redressed. ($1,500+). *Courtesy of Paula Reding and the Denver Doll Emporium.*

12" tall French bisque head pouty character girl by S.F.B.J. She has a bisque socket head, brown sleep eyes, a pouty expression, and a mohair wig. The doll has a French wood and composition fully jointed toddler body with painted shoes and socks. Her original S.F.B.J. sticker is on the back of her torso. Her clothing appears to be original. She is marked "S.F.B.J./252 Paris 4". ($3,500-4,500). *Courtesy of Frasher's Doll Auctions, Inc., Oak Grove, Missouri.*

Adolf Wislizenus

The German Adolf Wislizenus firm marketed dolls with bisque heads on ball-jointed bodies. Their factory made the bodies and other companies, including Simon & Halbig and Ernst Heubach, produced the heads. The firm was responsible for several character dolls including one with the mold number 115. Most of their dolls were marked "A.W."

12" bisque socket head little boy doll, marked "115". He has molded hair and painted features including an open-closed mouth with teeth. The doll has a composition ball-jointed body. A similar doll marked "115" has been identified as having been marketed by the German Wislizenus firm. The company specialized in ball-jointed bodies and used heads made by other companies. He has been nicely redressed. ($600-800). *Denver Doll Emporium.*

Chapter Two

Celluloid Dolls

The exact date when celluloid dolls were first manufactured is not known. It *is* known, however, that the Hyatt Brothers of Newark, New Jersey were producing products made of celluloid as early as 1869 under the name Celluloid Novelty Co. According to the Colemans, writing in *The Collector's Encyclopedia of Dolls Vol. I*, William B. Carpenter received patents for celluloid dolls in 1880 and 1881. These patents were both assigned to the Celluloid Novelty Co.

Although celluloid doll production apparently began in the United States, Germany became the country that manufactured the largest quantity of dolls made of celluloid. This production began around 1896. The material was light in weight and therefore cheaper than the heavier bisque for Germany to use in producing dolls for export. The biggest problem with the celluloid material was the fire hazard associated with it. Several factories burned because of this danger and production of celluloid was banned in Germany during the 1960s.

Germany

The most prolific German manufacturer of celluloid dolls was the Rheinische Gummi und Celluloid Fabrik Co., founded in Bavaria in 1873. This firm used the tortoise (Schildkrot) mark on their products beginning in 1889 and, according to Angelika Lipinski writing for *Doll Reader* in January 1990, the first celluloid dolls were produced in 1896. The first dolls were simple designs with two halves of the stiff dolls joined together. Later, movable arms were added to make the dolls more realistic. The firm continued to improve its products by manufacturing celluloid shoulder heads, all celluloid dolls, arms and legs made of celluloid to be used on kid or cloth bodies, and celluloid socket type heads that fit on composition bodies.

The tortoise trademark changed during the years, according to Lipinski. From 1899 until 1914, products were marked with an unframed tortoise and usually the words "Germany" and "Schutz-Marke" (trademark). From 1914 to 1933 (except for the war years), the tortoise was enclosed in a diamond frame. After 1933, the tortoise that was enclosed in the diamond frame included legs slanting to the back.

The firm also produced blank dolls and special order celluloid dolls that were sold to other companies, including the German firm of Kämmer & Reinhardt during their many decades of business. Most of the heads were based on bisque heads sold by the firm. Some of these heads also carry the tortoise logo as well as the K & R trademark. Mold marks for celluloid heads include #715, 717, 728, and 789 (the last mark is found on a walking doll). These are all pictured in the Kämmer & Reinhardt 1927 catalog. Earlier shoulder heads include #225 and #255. Other famous K & R bisque dolls also produced in celluloid include Marie and Peter (#101) and the so-called Kaiser baby #100.

The J.D. Kestner firm also contracted with the Gummi und Celluloid Fabrik Co. to produce their celluloid heads. The tortoise mark as well as the Kestner name usually appears on the dolls' heads.

In later years, Gummi und Celluloid Fabrik Co. produced celluloid dolls for companies in the United States, including Bonnie Babe, Sunny Girl and Boy for Averill, and Baby Bo-Kaye for Joseph Kallus.

In 1929 the firm was sold to the IG-Farben-Group, who continued to produce dolls.

The German firm of Buschow & Beck, founded in 1890 and more famous for its metal head dolls, also produced celluloid dolls for many years. The company used their Minerva trademark for these dolls as well. The early dolls usually had "Minerva" plus the helmet design stamped on the front of the doll, along with "Schutz-Marke" (trademark) and the word "Germany" on the neck or back of the head. "Schutz-Marke" was discontinued as a marking after 1925. According to Angelika Lipinski writing in *Doll Reader* in September 1992, the Buschow & Beck firm came up with the idea of covering tin doll heads with a layer of celluloid to make them more realistic. The firm marketed the first doll using this technique in 1903 and continued to produce this type of doll for decades. The company eventually produced celluloid shoulder heads, all celluloid dolls (in both little girl and baby styles), and celluloid socket heads. The company went out of business in 1950.

Other German firms who marketed celluloid dolls included Schoberl & Becker (marked with a Mermaid design) under the Cellba trade name and Kämmer & Reinhardt, and J.D. Kestner. The Karl Standfuss company produced celluloid doll heads plus celluloid dolls and parts in the early years. By the 1920s, they were manufacturing celluloid Kewpies and Bye-Lo Babies. They sometimes marked their dolls "Juno."

20" tall Kämmer & Reinhardt celluloid shoulder head doll featuring a head made by the Rheinische Gummi und Celluloid Fabrik Co. for K & R before 1914. She has glass sleep eyes, real eyelashes, an open mouth with teeth, and her original real hair wig. Her lower arms and legs are celluloid connected to a jointed leather body. The back of her shoulder plate is marked "K (star symbol) R/255." The doll is also marked with the old tortoise design with no diamond frame around it, which dates the doll to 1914 or before. She wears lovely old clothing but it is not original. ($300+).

23" tall Kämmer & Reinhardt doll with a celluloid socket head mounted on a five-piece composition toddler body, circa mid to late 1920s. She has flirty sleep eyes, open mouth with teeth, and her original mohair wig. She is marked on the back of her head "K (star symbol) R/728/10." Similar dolls were still being featured in the FAO Schwarz Christmas catalog in 1930. These dolls had celluloid heads, flirting eyes, toddler bodies, and wigs. They were dressed in dotted Swiss dresses, came in sizes from 12" to 18" tall, and were priced from $10.00 to $16.00 each. This doll has been redressed. ($350+).

The head of the K-R 728 toddler includes flirty eyes and eyelashes as well as an open mouth with teeth.

13" tall J.D. Kestner celluloid shoulder head doll with a head made by the Rheinische Gummiund Co. before 1914. She has glass eyes, open mouth with teeth, and a replaced wig. Her body and legs are leather with lower bisque arms. The shoulder plate is marked "J.D.K./201/3/0." The markings also include the early Rheinische tortoise with no frame, indicating that the doll head was made by that firm between 1899 and 1914. She has been redressed. ($150+).

Left:
24" tall celluloid shoulder head doll made by the German Rheinische Gummi und Celluloid Fabrik Co., circa 1915-1920. She has glass sleep eyes, open mouth with teeth, and her original mohair wig. Her lower arms and legs are celluloid attached to a leather jointed body. She is marked on the back of her shoulder plate "16 ½" along with "Germany/Schutz-Marke" and the firm's diamond framed tortoise. She has been redressed. ($250+).

18" similar celluloid shoulder head doll in a smaller size, probably made by the same German firm. Her blue eyes are especially attractive. She has her original mohair wig plus open mouth and teeth. ($175).

16" tall celluloid shoulder head doll made by the Rheinische Gummi und Celluloid Fabrik Co., circa 1920s. She has sleep eyes, open mouth with teeth, and her original mohair wig. Her lower arms are celluloid and her body, upper arms, and legs are cloth. She wears her original underwear, shoes, and socks. Her only mark is a framed tortoise, which was in use until 1930. ($200+).

14" tall celluloid shoulder head doll made by the Rheinische Gummi und Celluloid Fabrik Co., circa late 1920s. She has painted eyes, open mouth with teeth, and what appears to be her original wig. Her lower arms are celluloid while her body and legs are a leather type material. Her clothing (except shoes) appears to be original. She is marked with the framed tortoise. ($125)

16" tall celluloid shoulder head doll made by the Rheinische etc. company, circa 1920s. He has glass eyes, molded hair, open-closed mouth with teeth, cloth body and upper legs, composition lower legs and full arms. The back of his shoulder plate is marked with the diamond framed tortoise as well as "Germany 15." He looks very much like the Madame Hendren Sunny boy (and girl) dolls circa 1927. The heads for those dolls were also produced by the same German firm. He has been redressed. ($200+).

8" tall, all celluloid girl doll made by the German Buschow & Beck firm, circa 1930. She has molded hair, painted eyes, and a closed mouth. She is jointed at the shoulders and hips. She is marked "Minerva" along with the firm's helmet trademark design and "Germany/18/20." She may have originally been dressed in a regional costume. The 6" all celluloid baby was also produced by Buschow & Beck. He has molded hair, painted features, and open nostrils. He is jointed at the shoulder and hips. This doll is marked on the back "Minerva" followed by the helmet trademark along with "15/Germany/0/15 ½". Both dolls have been redressed. The metal buggy is circa 1930s. (Girl $50+, baby $40).

10" tall, all celluloid little girl doll by the Rheinische etc. firm, circa 1930. She has molded hair, painted eyes, and a closed mouth. She is jointed at the shoulders and hips. Her back carries the tortoise trademark. She was dressed in hand made clothes many years ago. ($50).

11.5" tall, all celluloid little girl doll made by the same firm, also circa 1930. She has molded hair, painted eyes, closed mouth, and is jointed at the shoulders and hips. The trademark tortoise in the diamond frame is on her back along with "29". She has been redressed in a dress from the period. ($75+).

7" and 8" tall pair of celluloid shoulder head dolls made by Buschow & Beck circa 1930s. Both dolls have molded hair, painted features, and cloth bodies. The dolls are all original and were probably sold as souvenirs for the tourist trade. The vintage tag on the boy's arm reads "Swiss German border." Much of the clothing and the arms and shoes are made of felt. The heads are marked with the firm's helmet trademark as well as "0 Germany 6 ¼". ($50 pair).

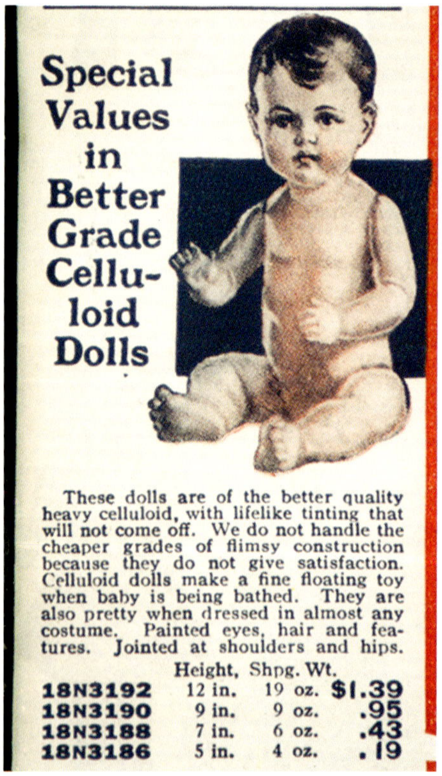

Larger celluloid baby dolls were also produced by Germany for many years. This one was advertised in the Fall and Winter Sears catalog of 1923. It came in sizes of 5" to 12" tall, priced from 19 cents to $1.39 each. The dolls had painted hair and eyes. The advertisement suggested that they could also be used as bath toys.

12" tall celluloid shoulder head doll marked "Germany" on her back. She dates from 1920 or before. The doll has stationary glass eyes, an open mouth with teeth, her original mohair wig, lower bisque arms, and a leather jointed body with cloth lower legs. The lower legs are black to represent socks. She has been nicely redressed. ($125+).

5" tall, all celluloid doll marked with a star on her back. She appears to be a German product circa late 1920s. She has molded hair with a molded hair bow, painted features, and is jointed at the shoulders and hips. ($35+).

11" tall German celluloid shoulder head doll, circa 1920s. She has molded hair and painted features. Her lower arms are bisque. The doll is completely original including her perfect shoes. She may be marked under her costume but her clothing has never been removed. ($175+). *Doll from the collection of Ellyn McCorkell.*

Newer German all celluloid dolls, circa 1940s-1950s. The larger 19" tall girl doll has molded hair, glass eyes, and a closed mouth. She is jointed at the neck, shoulders, and hips. She is marked on the back with the mermaid symbol and the Cellba (Schoberl & Becker) trademark. The 16" jointed baby doll has molded hair, brown glass eyes, and an open mouth. He is jointed at the neck, shoulders, and hips. He is marked with the newer tortoise mark with the legs slanted backward. Both dolls have been redressed. Celluloid dolls were no longer made after it was learned that they were a fire hazard. ($100+ each).

France

Doll companies based in France were also active in the celluloid doll trade. According to Dominique Penneques writing in the July 1991 *Doll Reader*, one of the earliest producers of French celluloid dolls was the Anel Co.

The Anel firm produced an all celluloid baby that was featured in the company catalog in 1913. Other products included a walking doll with a celluloid head and a black bent leg baby made of all celluloid. This company ceased doll production in 1928.

Another famous French celluloid doll distributor was the Societe Nobel Francaise, founded in 1927. According to Penneques writing for *Doll Reader* in November 1991, it united several earlier firms that had been working in the celluloid doll trade. Most of these dolls are marked "SNF" in a diamond frame. Included in their inventory were all celluloid baby dolls, a black Jacky doll, and all celluloid dolls with both molded hair and wigs. Various models of dolls in an assortment of sizes up to 27.5" tall were produced. The firm is no longer in business.

Petitcollin, another French firm responsible for very interesting celluloid dolls, was founded in 1856. According to Penneques writing in *Doll Reader* in October 1991, the firm began producing celluloid dolls in 1909. The company marked their products with a "head of the eagle" as their trademark. They produced all celluloid dolls (both children and babies) and celluloid shoulder heads with both painted and glass eyes and with either molded hair or mohair wigs. In 1930, Petitcollin merged with another firm and patented a new doll called Petitcollin. The doll continued to be made over several decades with glass stationary or sleep eyes, and with straight or bent legs. He was jointed at the neck, shoulders, and hips. The Penneques information indicated the new firm continued to produce celluloid dolls until 1957.

12" tall celluloid shoulder head boy made in France by Petitcollin, circa 1920. He has molded hair, painted eyes, and an open mouth with teeth. His unusual body includes lower bisque arms on a cloth body similar to leather bodies of the period. He is jointed at the shoulders and hips. He is marked on his shoulders "France" along with an Eagle head motif and the number 2. He has been redressed. ($125-150).

Poland

The country of Poland also was home to a firm that manufactured celluloid dolls circa 1920s-1930s. They were produced by P.R. Zast. The letters "ASK" were also included in markings on these dolls.

20" tall celluloid shoulder head doll made in Poland. She has molded hair, painted features, celluloid hands, and a cloth body. Her shoulder plate is marked "P.R. ZAST" with a triangle under the company name that includes the letters "ASK." Although the doll has been redressed, her original clothing consisted of a regional costume with a short skirt. Age unknown. (Redressed $50-75).

Japan

During the time of broken trade relations with Germany, i.e., before, during, and after World War I, American consumers were unable to purchase toys that had previously been imported from Germany. As a result, manufacturers in Japan began producing celluloid and bisque dolls to fill this need. Because of the success of the products and the cheaper prices of the celluloid toys, Japan continued to offer celluloid exports in the toy and doll line until World War II. Most of the celluloid dolls are marked "Made in Japan" or "Japan," so they are easy to identify. The majority are of lesser quality than their German counterparts but some of the nicer examples compete well with the German products. The particular firms that produced the dolls are not known. Japanese novelty dolls such as the Kewpies, carnival dolls, football players, and dolls with molded clothes are fun to include in a collection that offers a wide variety of dolls.

5" and 8" pair of all celluloid dolls made in Japan, circa late 1920s, early 1930s. The dolls need to be restrung. They have molded hair, painted features, and molded clothing. Both dolls are jointed at the shoulders and hips. ($15-30 each).

Pair of all celluloid dolls made in Japan. The doll on the left is 14" tall and has molded hair, decal type eyes, and is jointed at the shoulders and hips. He is marked on the back with a cross in a circle and "Japan." The other 15" baby doll is made of a lightweight celluloid. She has molded hair, painted eyes, and is also jointed at the shoulders and hips. She is marked on the back but not clearly enough to be read. She was probably also made in Japan. Both dolls have been redressed. (Boy $60+, girl $45+).

United States

The most famous of the celluloid doll producers in the United States was the Parsons Jackson Co. of Cleveland, Ohio. According to the Colemans in their book, *The Collector's Encyclopedia of Dolls*, the firm was known as the Parsons & Parsons Co. in 1918, and the dolls are so marked. At this time, the location of the firm was listed as New York City. The dolls were first placed on the market in 1910 and were advertised as being made of a material called Biskoline (which was a heavier celluloid type material). They were marked with a stork on the back along with "Trademark/Parsons-Jackson Co. Cleveland, Ohio." The baby dolls were assembled with steel springs and came in sizes from 10.5" to 14", priced at $1.50 to $2.50 each.

These dolls can still be found in very good condition because they did not crack or break, however the rather plain dolls do not seem to appeal to today's collectors. Since many of the dolls are nearly one hundred years old, they are still in remarkable shape.

Many other companies in the United States produced or imported celluloid dolls, but there is little information on most of these firms. The Viscoloid Company, located in Leominster, Massachusetts, is the exception. It was in business from approximately 1901-1930s. According to the book *Celluloid Dolls, Toys and Playthings* by Julie Robinson, the company was purchased in 1925 and became the DuPont Viscoloid Co. The firm produced dolls as well as many other celluloid toys through the years. The dolls are marked with a large "V" with "USA" across the top of the "V" along with a large "C" horizontally displayed across the bottom of the "V" with an "O" enclosed by the "V" and "C".

Several celluloid dolls made from German molds and marked "U.S.A." have been found. Perhaps they were produced during and after World War I, when trade relations with Germany had been broken and no laws of infringement had been broken. Celluloid dolls marked with a full Eagle figure have also been seen. These may be United States products but the company using that mark is not known.

Marks Brothers, from Boston, Massachusetts, used a shield containing the words "Made in U.S.A. Marks/Brothers/Co./Boston" as a trademark for their celluloid dolls. They were in business beginning in 1918 and into the 1930s, according to the Colemans writing in their *Collector's Encyclopedia of Dolls, Vol. II*. A raised Indian head and "American" were used on other celluloid dolls made by an unknown American firm.

12" tall celluloid shoulder head doll made in the U.S.A., circa 1920. He has molded hair, painted features, (including teeth), and lower celluloid arms. His leather body has lower legs made of cloth. He is marked "2/Made in U.S.A." the head may have been made from an earlier German mold. ($100+).

Below:
6" tall, all celluloid baby doll made in the United States circa 1920s. She has a replaced wig (round hole in head for wig), painted features, and is jointed at the shoulders and hips. A full Eagle motif with wings trademark is on her back. This mark was used by an unknown company in the United states. She has been redressed ($35).

4.5" tall pair of all celluloid dolls dressed in Pilgrim costumes made of crepe paper. The dolls are pictured with their original circa 1920s box marked "A. S. Burbank/ Pilgrims Book and Art Shop/Plymouth, Mass." Picture on front, labeled "Canopy Over Plymouth Rock." The dolls were sold to tourists as souvenirs. They have molded hair, painted features, movable arms, and probably stationary legs. The clothing has been glued to their bodies. (MIB pair $75).

Pair of all celluloid baby dolls thought to have both been made in the United States. The 10" all celluloid doll on the left has painted features and hair, closed mouth, and is jointed at the shoulders and hips. The doll is marked on the back with an Eagle (full bird), which indicates it was probably made in the United States. The German Rheinische Gummi und Celluloid Fabrik Co. made a doll exactly like this one (even to the painted hair), circa 1925. The 11" doll on the right was made from a similar mold. It is marked with an intertwining CVO design along with U.S.A. The doll was produced by the Viscoloid Co., located in Leominster, Massachusetts. This appears to be an earlier wigged model. The skin tone on this doll has mellowed to ivory. Both dolls have been redressed. ($100 each.)

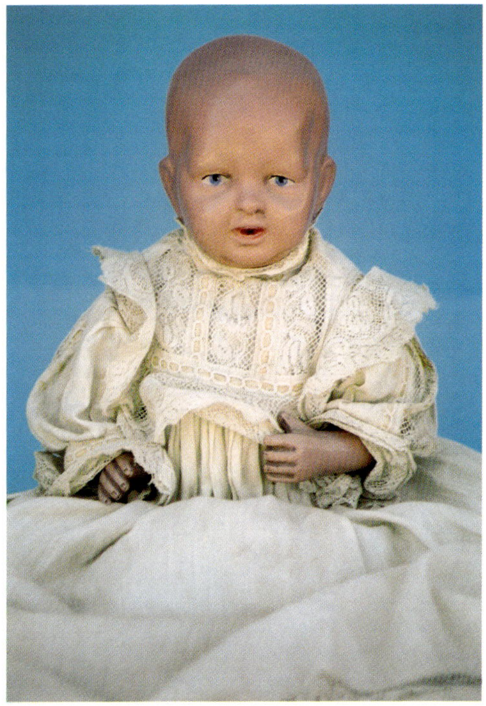

12" baby doll made by the Parsons-Jackson Co. in Cleveland, Ohio, circa 1912-1917. The dolls were made of a material the firm called Biskoline. Although these dolls are usually classified under the celluloid category, the Biskoline material is far sturdier than the usual celluloid used for doll production. The Parsons-Jackson dolls do not crack or break as do the regular celluloid dolls. The doll has painted features and spring jointed arms and legs. It is marked with a stork on the back of the head and another stork on the back of the body with the words "Trade Mark." Lower on the back is "Parsons-Jackson Co./Cleveland, Ohio/U.S.A."

19" tall Pinky Winky Doll created by Madame Georgene Averill, circa 1924. The doll was advertised in *Playthings* magazine in September of that year by the "Pinky Winky Products Co., Inc." of Boston, Massachusetts. The doll's face and front of the head and shoulders are made of a celluloid type material called Fiberoid. The rest of the doll, including the arms, legs, and back of the head, are made of cloth. The back of the cloth head has hair printed on it. The doll has a cry box in her back. The black pieces in the Google eyes are supposed to be in vertical positions. In the ad, the doll said, "I roll my eyes and say Ma-Ma." The doll sold for $3.75. She has been redressed. (In this condition $75+).

Celluloid dolls were at their height of popularity from 1910 to 1940, and even though the dolls are very breakable, many fine examples have stood the test of time very well. Collectors need to treat these dolls with special care, however, particularly when placing a celluloid doll in the mail. Just wrapping bubble wrap around the doll and placing her in a box does not insure safe delivery. If the bubble wrap is taped around the doll too tightly, she may break or crack due to the outside wrapping. The dolls should always be double boxed for transportation, with good packing materials encasing the inside box. The older dolls are especially fragile. There has been a current surge of interest in the old celluloid dolls that had been lacking for many years. Perhaps with this new appeal, more of these treasures will be carefully preserved for future generations.

Chapter Three

Cloth Dolls

The years from 1910 to 1930 provided great cloth dolls from many companies based in several different countries. Firms from the United States offered printed dolls like those made by the Art Fabric Mills and Arnold Print Works. The Babyland and Bruckner dolls offered a new concept of "mask" face dolls, while the Alabama, Chase, Rollinson, and Kamkins dolls featured more realistic cloth dolls with molded faces. The most popular cloth dolls from the period were the Raggedy Ann and Andy dolls, based on characters in the books by Johnny Gruelle. These dolls are still being sold today.

English manufacturers offered a series of more art-like dolls during this period. Included were the Chad Valley, Norah Wellings, and Dean's Rag dolls. Most of these used felt and/or velvet in their construction or clothing.

The original German Käthe Kruse dolls were also from this period. Even these dolls have been re-invented over the years, using different materials like celluloid. The oldest dolls are still the most desirable for a collector. The German Steiff dolls as well as animals were becoming popular toys for children in many countries during this time. Steiff Teddy Bears from any era remain a stable collectible for both children and adults.

It's possible that the wonderful old Lenci dolls from Italy remain as the most collectible of cloth dolls today. These dolls were reproduced in recent years and collectors who can't afford the expensive originals have an opportunity to purchase one of the newer models. These dolls were also very well made.

Many other companies of the era also produced "Art Dolls" in the Lenci manner to take advantage of the Lenci dolls' popularity. The Raynal dolls from France are one example. They, too, usually used felt in the production of their dolls and for clothing.

Another interesting field of cloth dolls are those made to advertise a variety of products. Kelloggs and Aunt Jemima pancakes were just two of the many firms that used this method to promote their products. Most of these dolls were printed on cloth and were meant to be cut out and stitched together by the owner in order to make a flat-faced printed doll.

With so many cloth dolls to choose from, some collectors concentrate on purchasing dolls from one firm—Lenci or Raggedy Ann for example—and building a whole collection with only these dolls. Other collectors prefer a certain type of doll and include Alabama, Chase, Rollison, and similar dolls in a collection. Still others may want to build a collection that features only advertising dolls. Most collectors, however, just add a few cloth dolls to a more versatile collection that includes bisque, china, and a variety of doll types. Regardless of what method is used, the wonderful array of cloth dolls from this period should not be overlooked.

United States

Alabama Indestructible Dolls

Ella Smith of Roanoke, Alabama was issued a patent on the Alabama Indestructible Dolls in 1900. According to Jacque Shafer writing in the *Antique Doll Collector* in June 2002, The dolls were made by using a wet knit fabric that was pressed into molds. The molds were later filled with plaster until dry. Afterwards, the cloth dolls were finished with oil paint. Some of the dolls were molded with bare feet while others had painted shoes and socks. The early dolls had applied ears and painted hair. Later dolls featured molded ears and sometimes wigs as well as painted hair. All of the dolls had molded faces. Some of the Alabama dolls were stamped on the front of the body "The Ella Smith/Doll Co. Made in U.S.A./The Alabama Indestructible Doll/Pat. Nov. 1919." Dolls were made in sizes from 11" to 24" tall.

Author Shafer states in her article (titled "Ella Smith Doll Manufacturer") that for a time, circa 1919, the dolls were produced in a small doll factory until it closed in 1922.

14" tall American later "Alabama Indestructible Doll" made by Ella Smith in Roanoke, Alabama, circa mid-teens. The cloth doll was painted with oils (including her hair and features). She also has painted shoes and socks. Her ears are molded instead of applied, which indicates she is a later version of the Alabama doll. ($1,400-1,600). *Photograph and doll from the collection of Jean M. Grout*

Arnold Print Works

The Arnold Print Works was in business in North Adams, Massachusetts from 1876 until at least the 1920s. The firm's doll products were made from printed fabrics. The dolls were to be cut out, sewed together, and then stuffed. Usually they were made in two pieces, a front and a back. According to the Colemans, writing in *The Collector's Encyclopedia of Dolls*, the fabric sold for 12 cents to 20 cents a yard but many dolls were printed on only a half yard of material so these toys were very inexpensive.

Some of the company's popular dolls included the Palmer Cox Brownies, Soldiers, and a 1912 Gibson Doll that was jointed and came with directions to make the doll. According to Johana Gast Anderton writing in her book, *The Collector's Encyclopedia of Cloth Dolls*, the Toy Works of Middle Falls, New York reproduced many of the Arnold Print Works dolls in the 1970s. Collectors should be aware that some of these dolls have been represented as being old by unknowing sellers. Included are Blossom, (originally Pickaninny) as well as other examples taken from original items in the toy collection of the Museum of the City of New York.

16" tall cloth Arnold Print Works Columbian "Sailor Boy" as pictured in the ad from *The Youth's Companion*. The country had been celebrating the 400th anniversary of Columbus' discovery of America and the copy in the ad stated that this doll, which was patented in 1893, would be a good souvenir of the celebration. He is marked on the back of his head "Arnold Print Works. PAT. Jan. 31, 1893." He was originally printed on a half yard of material, then sewed together and stuffed at home. ($75).

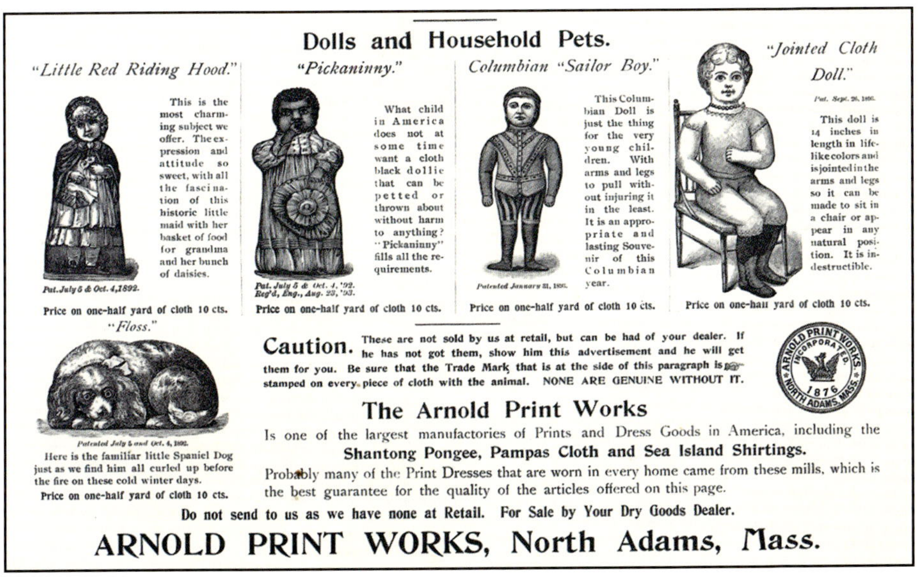

Arnold Print Works advertisement from *The Youth's Companion*, dated November 30, 1893. Although this advertisement is quite old, many of the same animals and dolls were made for many years. The toys were printed on fabric and were to be sewed and stuffed at home. Pictured are "Little Red Riding Hood, Pickaninny, Columbian Sailor Boy, and Jointed Cloth Doll. Each of the dolls was printed on a half yard of cloth and sold for ten cents.

The Art Fabric Mills

The Art Fabric Mills was located in New York City, where it began business in 1899. A patent had been received by Edgar G. Newell in 1900 for a life-sized cloth doll. These dolls were flat faced little girls, made in several pieces, which could be assembled at home. The doll was reproduced on sateen from an original hand-painted doll. Changes were made to the original doll over the years, such as more realistic feet including printed shoes that protruded in front. The dolls ranged in size from 20" to 32" tall. Later dolls were made of drill cloth.

Other dolls produced by the firm included George and Martha Washington dolls, (measuring nearly two feet tall) comic strip related Foxy Grandpa, and the Newlywed Kid. The firm went out of business circa 1910.

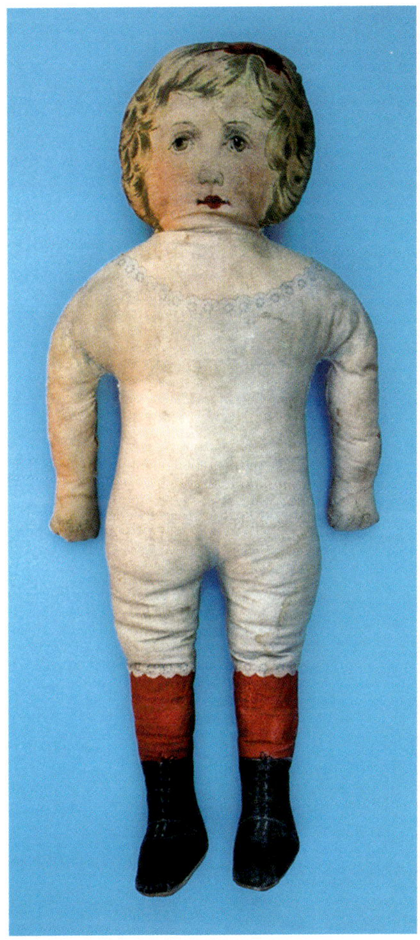

26" tall cloth doll made by the Art Fabric Mills of New York. The firm produced many different fabric dolls over the years. This one has the new patented feet that were supposed to allow the doll to stand. She also has printed underwear, socks, and boots to dress up the doll. She was to be sewed and stuffed at home. The bottom of the foot is marked "ART FABRIC MILLS/New York/ patented Feb. 13th 1900." The doll was intended to be large enough that she could be dressed in baby cl othes. ($125-150 with some wear).

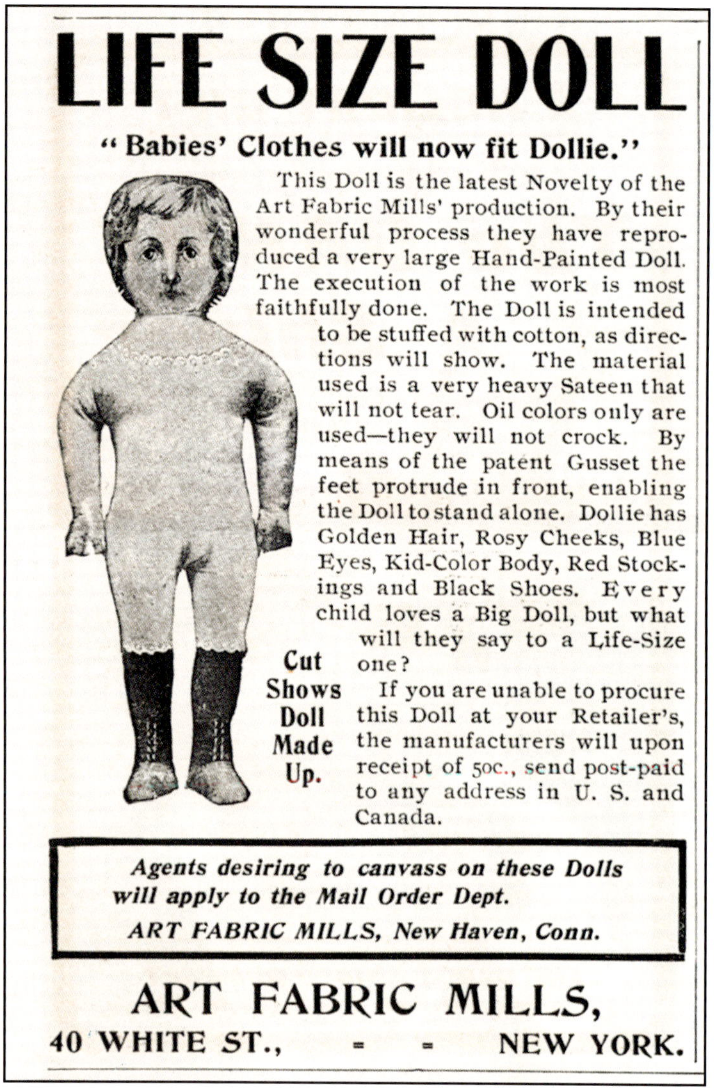

Art Fabric Mills advertisement from *Munsey's Magazine* of December 1901. The New York firm stated that the doll was made of heavy sateen "that will not tear." Directions were included so the consumer could assemble the doll and stuff it with cotton at home.

Babyland Rag-Horsman-Bruckner

The E.I. Horsman firm of New York City included "Babyland Rag Dolls" in their catalogs from 1893 to the 1920s, according to Don Jensen writing in his book, *Collector's Guide to Horsman Dolls 1865-1950*. The earliest dolls had flat, hand painted faces. In 1907, according to Jensen, photographic faces were added to some dolls.

The Horsman Babyland catalog circa 1912 offers yet a third type of face for the cloth dolls. These dolls, made by the Albert Bruckner firm in New Jersey, included mask molded faces instead of the earlier flat faces. The 1912 catalog features dolls of all three types in a variety of sizes. The Topsy Turvy doll (black head on one end, white on the other) was pictured with the photographic faces. The Bruckner firm may have also produced these dolls. Twenty-nine different Babyland dolls were pictured in what appears to be the 1912 Horsman catalog. They ranged in size from 13" to 30" and were priced from 50 cents to $5.00 each.

The Bruckner Topsy Turvy dolls with the mask molded faces were still being sold as late as 1928. The Bruckner firm was then known as Albert Bruckner's Sons, located in New York City. Other more modern cloth dolls carried by the firm at this time included Dolly Pop, Tubby-Tot, Little Brother, and Red Riding Hood.

14" tall black Babyland Rag doll with the flat hand painted face. She is all original. She probably retailed originally for $1.00. This black doll also has a mohair wig. ($1,000-1,200). *Grout Collection.*

30" tall Babyland Rag doll with a flat oil-painted face. She wears her original clothing. A similar doll wearing a dress like the one on this doll was featured in a "Babyland Rag Dolls" catalog issued by the E. I. Horsman Co., circa 1912. The cost of the pictured 30" doll was $5.00. These early dolls had hand painted faces and were jointed at the shoulders and hips. Some examples, like the one pictured, had mohair wigs. ($2,800-3,200). *Photograph and doll courtesy of Jean M. Grout.*

14" tall Babyland doll on the left with the "New Lifelike Face." These dolls featured photo lithographed faces. ($800-900). Besides the 30" tall doll already pictured, a 14" topsy turvy black and white doll is also shown. Although she was part of the Babyland line, she was probably made by the Albert Bruckner firm. These dolls have molded faces instead of flat faces, as on the earlier Babyland dolls. *Grout Collection.*

13" tall Bruckner's "Tu-In-One" doll, which featured a black doll on one end and a white doll on the other. The doll is marked with a tag reading "Bruckner Doll/Made in U.S.A." The long skirt covers the doll's head and hands when not in use. The doll has molded mask faces like those the firm made for Horsman in earlier years. ($350+).

12" tall (including original dress 15") Babyland Rag baby doll. She is similar to one featured in an E. I. Horsman Co. catalog priced at 50 cents. She has a molded mask type face and wears all her original clothing. Both her cloth body and clothing were circa 1912 and were made very cheaply of inexpensive materials. Her hands do not have a thumb as do most of the Babyland dolls. Her bonnet ties of inexpensive twill have never been cut and her clothing has never been removed. She was found wearing a dress made over her original clothing that had been sewed directly to the doll. An additional bonnet had also been made over her original commercial one. "Patented" is printed on the right front side of her neck. ($300+).

13" tall cloth "Dolly Pop" type doll made by Albert Bruckner's Sons of New York City, circa late 1920s. This is the boy version. The doll came in a variety of costumes as pictured in the Bruckner Dolls ad from 1928. He is missing his sailor hat and shoes and socks. ($200-250). *Photograph and doll courtesy of Jean M. Grout.*

Albert Bruckner Dolls advertisement from *Junior Home Magazine* of December 1928. The New York City based firm had been producing cloth dolls since at least 1901 when it received a patent for a special doll face that was then used on their dolls. Many of these dolls were sold by Horsman as Babyland Rag Dolls. This advertisement pictures cloth dolls developed by the firm. Included are Tubby-Tot (13" $1.25), Red Riding Hood (15" $1.75), Every Little Girl Wants a Little Brother (15" $1.75), and the popular Dolly Pop Doll (13" $1.25). The firm was also still selling their "Tu-In-One reversible Negro and White" doll that had been a part of their doll inventory for decades. It sold for $1.75.

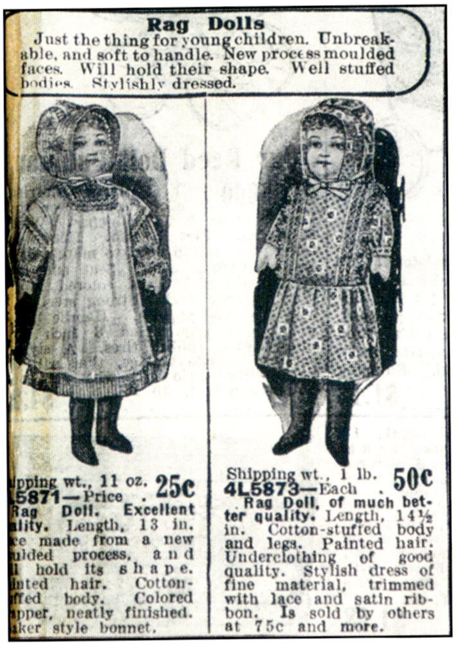

Rag doll advertisement in the Montgomery Ward 1914 catalog. The copy reads "New process moulded faces. Will hold their shape." The dolls sold for 25 cents and 50 cents each. The copy under the 50 cent doll reads "Is sold by others at 75 cents and more." The dolls are 13" and 14.5" tall. It is not known if the dolls were made by the Albert Bruckner firm. *Advertisement from the collection of Marge Meisinger.*

Chase Stockinet Dolls

According to Janet Johl, writing in her book *The Fascinating Story of Dolls*, Martha Chase of Pawtucket, Rhode Island, began making cloth dolls around 1889 for her own children as well as for neighborhood youngsters. In 1893, according to company information in 1941, she took one of the dolls to a Boston department store to buy some shoes. The doll was seen by store personnel and she received orders for the doll to be ready for Christmas of that year. The doll business grew from that point.

According to early advertising, the dolls were made of specially woven stockinet, stuffed with cotton. They were painted with oil paints and advertised as being waterproof. The ad stated that the features of the dolls were raised, formed, and hardened.

The earlier dolls had bodies of sateen and were jointed at the shoulders, elbows, hips, and knees. Later dolls from the 1920s were jointed only at the shoulders and hips. In 1909, dolls were selling at the Macy department store for $2.49 to $4.96 each, in sizes of 16", 17", 21", and 24". The dolls were marked under the arms or on the thigh with the Chase trademark (circle with face with the words "Chase/Stockinet Doll/Trademark." In 1910, dolls were being produced in sizes from 12" to 30" tall.

The Chase Hospital dolls were first produced in 1910-1911 as the result of a request by a hospital. These dolls continued to be used to teach nursing skills by hospitals worldwide until the 1960s or 1970s.

The early play dolls were dressed as either boys, girls, or babies. By 1921, dolls were also produced to represent characters from *Alice in Wonderland*, the Joel Chandler Harris stories, and from Charles Dickens novels. A George Washington doll was also produced. Since these dolls were made in smaller numbers, they are hard to find.

In 1922, girl dolls were made with Dutch cut bobbed hair. These dolls are also especially collectible.

Martha Chase died in 1928, but her family continued the business for many years. The new dolls that were produced into the 1970s were mostly used by schools and hospitals. They were made of vinyl coated cloth and included 16" boy and girl dolls and a 13" baby. The hospital babies were also still being produced during these years.

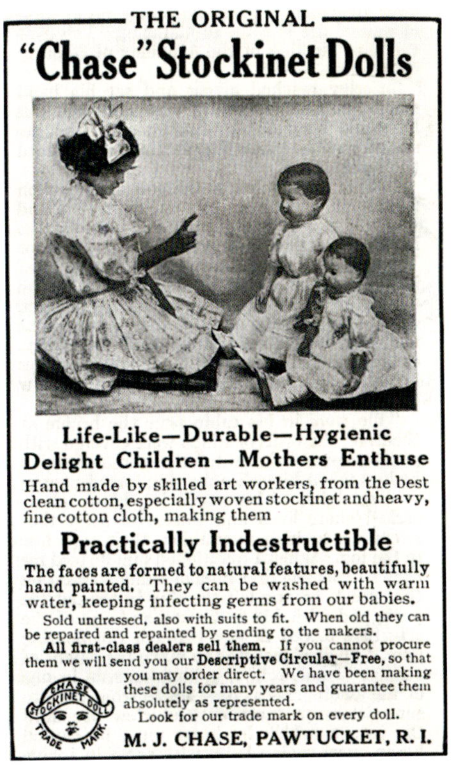

Chase Stockinet dolls were advertised in two ads from the *Ladies Home Journal* magazine in October 1911 (left) and November 1915 (right). The cloth dolls were made by Martha Chase of Pawtucket, Rhode Island, beginning around 1889. The ads state "Made of specially woven stockinet and fine cotton cloth filled with best clear cotton. The faces are formed and colored to resemble natural features." The dolls were sold either dressed or undressed. Pictured are examples of dolls dressed as babies, little girls, and little boys.

20" tall cloth "Chase Stockinet Doll" made by Martha Chase in Pawtucket, Rhode Island, circa 1911-1915. The dress is original and appears to be similar to the one shown on the larger doll in the ad from 1911. The same early model of doll was used to represent both girls and boys as seen in the 1915 ad. It is true that little boys often wore dresses in earlier years but by 1911-1915 little boys were more often dressed in rompers of some type. ($850-1,000). *Photograph and doll courtesy of Jean M. Grout.*

16" tall Chase doll jointed at the shoulders, elbows, hips, and knees, which indicates he is an early doll. He has been redressed as a boy. (With wear $400-450).

12.5" tall Chase little girl doll with bobbed hair and bangs, circa 1922. This later doll is one that is a favorite of collectors because she is hard to find. ($1,200-1,300). *Grout Collection.*

12" tall Chase baby doll. Those made to be used as baby dolls had fatter faces and bodies than the other dolls. ($900-1,000). *Grout Collection.*

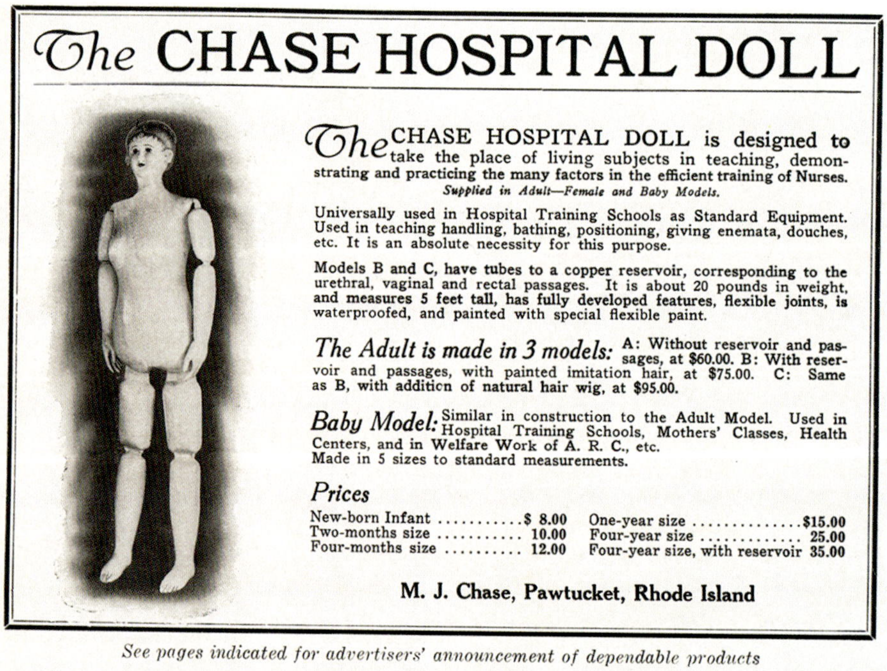

Chase Hospital Doll advertisement from 1927. Both the adult female and the various sizes of babies were to be used in nurses' training. The adult was five feet tall with flexible joints. They could be purchased with painted hair or wigs and with or without an inside reservoir. Prices ranged from $60 to $95. The infant models came in five sizes, from new-born infants to four year old children. These cost from $8 to $25 each, according to size. The four year old model also came with a reservoir for an extra $10. These dolls were made into the 1960s and 1970s of vinyl coated cloth. The older dolls are very collectible.

Kamkins Dolls

Louise R. Kampes was the designer of the cloth Kamkins dolls, which were produced from 1918 until 1934. Most of the dolls were from the Louise R. Kampes Studios at Atlantic City, New Jersey, in business from 1919 to 1928.

The Colemans, writing in *The Collector's Encyclopedia of Dolls,* describe the 1919 patent for the dolls as stating that unvulcanized rubber was used in production of the heads, which were covered by fabric.

According to Carol Corson writing in the *Antique Doll Collector* of October 2002, the dolls had molded mask faces with painted features and mohair wigs and were disc jointed at the neck, shoulders, and hips. They were produced in boy, girl, and baby models. The babies had bent limb baby bodies. The dolls sold for $10.00 to $15.00 each. The child dolls were 19" tall and the babies were 16" tall.

Many of the dolls were marked with a red paper heart reading "Kamkins/A Dolly Made to Love/Patented by L.R. Kampes Studios/Atlantic City/New Jersey."

By the mid 1920s, Corson states, the dolls were no longer disc jointed. The arms were attached by tabs and the legs were attached like most other cloth dolls, including the Georgene Novelty Dolls made in the 1930s.

Ads for 1934 gave the address for Kampes Dolls, Inc. at a showroom in New York City. Louise Kampes died in 1936 shortly after the demise of her dolls, circa 1934.

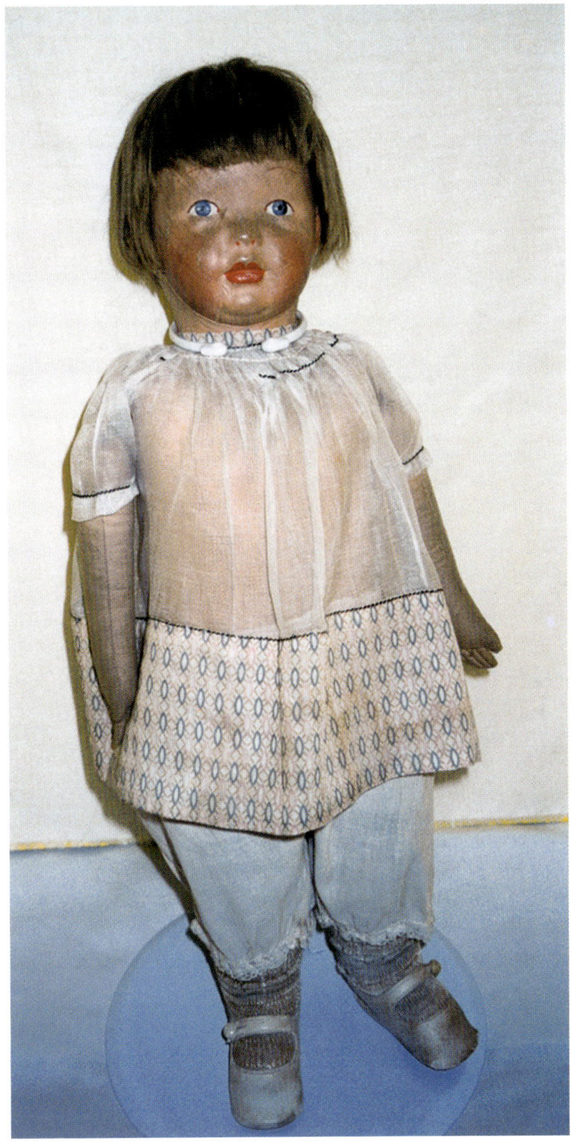

16.5" tall Kamkins little girl doll made by Louise R. Kampes, circa 1920s. She has a painted molded mask face that attaches to the stiffer back of the head. The body and the back of the head are made of peach colored cotton. The wigged doll has a red dot at the inside of both eyes and the same type dots indicate nostrils. Her eyes are blue and her head turns. The dress appears to be original. It is decorated with the familiar black silk floss used on most Kamkins dolls. The embroidery can be seen at the neck, sleeves, and hem. Originally, the matching bloomers probably showed just beneath the dress, but with the passage of time the elastic once holding them in place has allowed them to drop. The shoes and socks also appear to be original. Many of the dolls are marked with a red paper heart label on their chests reading "Kamkins/A Dolly Made To Love/Patented by L.R. Kampes Studios/Atlantic City, N. J." Others carry the same information with a stamp. ($1,000+). *Photograph and doll courtesy of Susie's Museum of Childhood at Bluebird Farms in Carrollton, Ohio.*

Nelke and Similar Dolls

The Nelke Corp. of Philadelphia, Pennsylvania, produced a series of cloth dolls for babies for over a decade, beginning circa 1917. The firm made both animals and dolls of a washable knitted cloth.

The Sears Fall and Winter catalog circa 1920 offered dolls 11" tall priced at 95 cents each. The dolls had painted eyes and features. Similar dolls were shown in Sears catalogs all through the 1920s and into the 1930s. A 12" tall corduroy covered doll sold for only 25 cents in the 1932 catalog. The price drop was probably not because of lesser interest in the product but as a result of the Great Depression.

According to Polly Judd writing in her book *Cloth Dolls*, 400,000 of the Nelke dolls were made in 1923.

Two different styles of Nelke cloth dolls were advertised in the Sears Roebuck catalog, circa 1920. The largest example was 14" tall with a painted face. He was dressed in a crepe romper suit with a knitted cap. The body was also made of knitted cloth, as were the other 11" tall dolls pictured in the same catalog. These dolls were made for babies and small children. The smaller dolls sold for 95 cents each while the 14" size was priced at $1.85.

10" tall cloth doll made by the Nelke Corp. (located in Philadelphia, Pennsylvania), circa 1920s. Soft dolls from this firm were advertised in the Sears catalogs during most of the 1920s. The dolls and animals provided especially good toys for babies because they were washable. This doll carries her original tag reading "This is a Genuine/ Nelke/Soft Toy/." ($150+). *Photograph and doll courtesy of Jean M. Grout.*

A similar 6" tall cloth Sweet Petiekin doll, evidently to be used as a baby toy. He is all original with a tag that reads "Sweet Petiekin/Trade Mark/ Reg. U. S. PAT OFF. Pat. May 13, 1924." ($275). *Grout Collection.*

Raggedy Ann and Andy

The Raggedy Ann cloth dolls have been a staple in the doll field since 1915, when they were patented by John B. Gruelle. The dolls were based on the characters from Gruelle's *Raggedy Ann Stories*, published by the P.F. Volland Co. of Joliet, Illinois. The same firm was also responsible for the manufacture of the dolls, at least from the 1920s until they went out of business in 1934.

According to Jan Foulke, writing in *Doll Reader* in September 1992, the earliest doll had brown hair with feet that turned out. Raggedy Andy didn't become a doll until after *Raggedy Andy Stories* was published in 1920. This book introduced the Andy character and the dolls have been successfully marketed as a pair over the decades since.

16" tall early Raggedy Ann doll on the right, thought to have been made by the P.F. Volland Co. of Joliet, Illinois circa 1920s, but she may be earlier. Her hair is brown instead of red since she is an early doll. Her clothing appears to be original. Her shoes are brown instead of black. The 16" Raggedy Andy is similar to an early Volland Andy doll that is known to have been made of white cloth. He has the large hands and printed features of those dolls but his hair is orange instead of dark red. The feet on these early dolls turn out and their lower legs are made of narrow red and white striped material. (Raggedy Ann $1,000). *Photograph and doll courtesy of Susie's Museum of Childhood at Bluebird Farm in Carrollton, Ohio.*

Rollinson Dolls

The Rollinson Dolls were designed by Gertrude F. Rollinson, circa 1916. They were similar to the Chase dolls and had molded oil painted faces and painted arms and legs. Some of the dolls had painted hair and others had added wigs. Most of the dolls had closed mouths but some included open-closed mouths with painted teeth. They were made of stockinette.

The dolls were produced by the Utley Co. of Holyoke, Massachusetts, according to information in *The Collector's Encyclopedia of Cloth Dolls* by Johana Gast Anderton. Dolls came in sizes from 14" to 28" tall. Some of the Rollinson dolls carried a stamp in the shape of a diamond, which included a doll in the center along with "Rollinson Doll Holyoke, Mass."

Anderson also pictures a Utley Fabco Doll from a *Playthings* ad in December 1917. These were all-fabric dolls jointed at the shoulders and hips. The dolls also had "beautiful eyes" and mohair or human hair wigs, according to the ad. These dolls no longer looked like Chase dolls. Their limbs and bodies were all made of the same materials and were jointed like a Lenci doll. The dolls came in sizes from 14" to 26" tall.

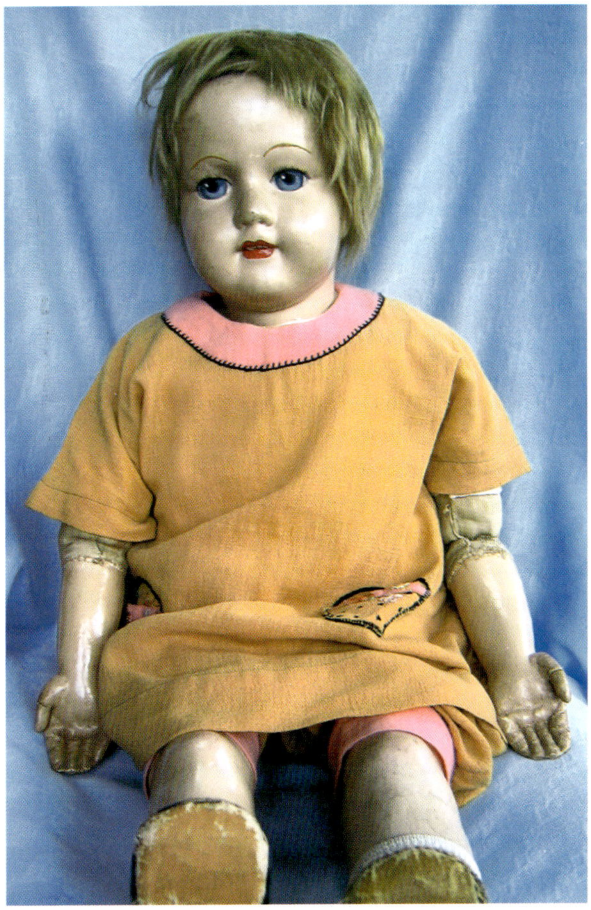

26" tall Rollinson wigged doll with open-closed mouth and painted teeth. She still retains part of the original red diamond stamp on her back. It originally read "Rollinson Doll Holyoke, Mass." ($1,400). *Grout Collection.*

16" tall Rollinson cloth dolls produced by the Utley Doll Co. in Holyoke, Massachusetts beginning in 1916. Gertrude F. Rollinson was the designer. The dolls with painted hair look similar to the Chase dolls from the same period. The dolls were made of molded cloth with painted heads and lower arms and legs. ($900-1,000 each). *Photograph and dolls courtesy of Jean M. Grout.*

Close-up of one of the Rollinson dolls with painted hair. The eyes include a large number of painted eyelashes, a red dot at the inside corners, and white dots in the eyes. *Grout Collection.*

Close-up of the cloth Rollinson wigged doll. The 26" size was apparently the largest size available. *Grout Collection.*

England

Chad Valley Co. Ltd.

The Chad Valley Co. Ltd. was an English firm located in Harborne, Staffordshire. It was founded by Joseph and Alfred Johnson circa 1920.

The earliest dolls were made of stockinet, with or without wigs. By 1923, blue glass eyes and mohair wigs were used on dolls made and/or dressed with velvet and plush fur. Most of the firm's dolls were marked "Hygienic Toys/Made in England By/Chad Valley Co. Ltd."

According to Polly Judd writing in her book, *Cloth Dolls*, the firm made many different designs of dolls throughout the 1920s. Included were Pierrot and Pierette, Long John Silver, Little Red Ridinghood, Jack and Jill, Jesters, Golliwogs, Peter Pan, and Boudoir Dolls. In addition, the firm offered a line of artist designed dolls by Mabel Lucie Attwell beginning in 1927. Attwell was an illustrator famous for her drawings of children.

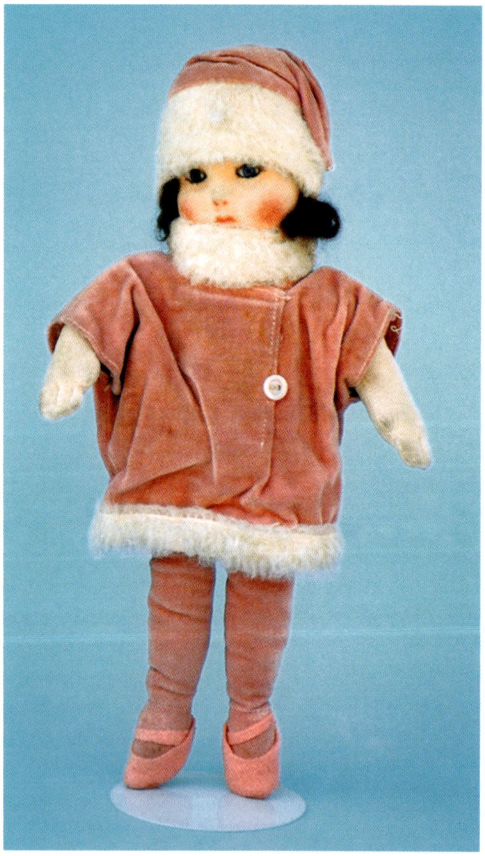

11" tall Chad Valley doll made in England, circa 1923. She has glass eyes, mohair wig, velvet arms and legs, and a cloth body. Her original pink velvet costume has white plush trim. She is unmarked but is similar to a labeled doll pictured in the *Cloth Dolls* book by Polly Judd on page 46. ($150+).

Dean's Rag Book Co.

Another English firm from this period was Dean's Rag Book Co., which had been established in 1903 in London. Their first dolls were printed on cloth and were to be cut out, sewed together, and then stuffed. These dolls were called "Cut-Out Knockabout Dolls." Beginning in 1980, the company reprinted many of these early dolls. Included in the reprints were Peggy and Teddy, which had been designed by Grace Drayton in 1912.

By 1920, according to author Polly Judd, more complicated dolls were being produced with molded pressed faces. This English firm also made a number of character and unique lines of dolls during the 1920s. Included were Charlie Chaplin, Buster Brown, and 21" tall Boudoir Dolls. One of their most unique ideas for a doll design was the "Dean's Dancing Dolls" model. This novelty item included two cloth dolls hooked together as if they were dancing. They debuted in 1928 and were an immediate success. Tourists were especially fond of these dolls and purchased them as souvenirs. The idea was also used by other firms in the United States who made dancing dolls with composition heads. Included were the Gerling Toy Co. of New York and the S.& H. Novelty Co. of Atlantic City, New Jersey.

Most of the Dean's Rag Dolls were marked on the bottom of the foot in an oval, "Hygienic A 1 Toys/Trade Mark/Made in England/Dean's Rag Book Co. Ltd." Two dogs, each pulling on one of Dean's indestructible books, were also part of the trademark.

The FAO Schwarz Christmas catalog for 1930 pictured a pair of dancing stockinet dolls 14" tall and priced at $8.50 for the pair. "A string suspension agitated by the hand supplied the motion power," according to the advertising. These dolls were probably made by Dean's Rag Doll Co.

12" tall, all original Dean's Dancing Rag Dolls made in England circa late 1920s. The dolls are marked on their feet "Hygienic A 1 Toys/Trade Mark/Made in England/Dean's Rag Book Co. Ltd." Two bulldogs playing with a Dean's book in an oval are also part of the marking. The cloth dolls have molded painted faces, mohair wigs, and cotton clothing (girl's clothing is trimmed in felt). The dolls were made to dance when dangled from a black string. A note came with the dolls, which reads "Bot by Dorothy Blatter in London in 1929." ($150-200).

Norah Wellings Co.

The Norah Wellings Co. was founded by Norah Wellings in 1926. The Victorian Toy Works factory, where the dolls were made, was located in Wellington, Shropshire England.

Ms. Wellings and her brother, Leonard Wellings, were partners in the enterprise. According to author Gillian Trotter writing in her book, *Norah Wellings: Cloth Dolls and Soft Toys*, the first dolls made by the firm were the Cora dolls. These dolls had felt faces, applied ears, velveteen jointed bodies, glass or painted eyes, and mohair wigs. They were dressed in voile, felt, and velveteen, and came in sizes of 13.5", 16", 19", and 28".

The 1929 catalog for the firm pictured twenty girl dolls and two boy dolls. There was also Miladye, a 22" tall boudoir type doll, and a baby doll called "Baby Mine," with glass eyes and painted hair.

Two of the most popular dolls made by the Wellings firm were also introduced in 1929. They were the Sailor dolls and the Black Islander dolls. The Sailor dolls came in sizes of 12", 14", and 18". The Black Islander Dolls came as both boy and girl models and had wide smiles and teeth, glass eyes, and wigs.

The Norah Wellings firm continued in business until 1959, when Leonard Wellings died. By that time, Norah Wellings was sixty-five years old so she closed the business and retired.

The dolls were tagged with a cloth tag that read "Made in England by Norah Wellings."

15" tall Norah Wellings "Norene" Dutch boy, circa early 1930s. He is all original and is wearing brown and orange checked pants with simulated wooden shoes made of felt. His jacket is lime green with brown and orange trim. His matching hat continues the brown and orange theme. These were unusual colors to use on a Dutch boy and probably represented quite a deviation from what would normally be found on a doll dressed in an ethnic costume. ($500-700). *Photograph and doll courtesy of Nancy Lazenby.*

France

Raynal

The Les Poupées Raynal dolls were produced in Paris, France, beginning circa 1922-1925. The firm was headed by Edouard Raynal.

These dolls were made of cloth and/or felt. They were jointed at the neck, arms, and legs. The heads were made of felt but, according to Evelyn Jane Coleman writing in *Doll Reader* in October 1989, the dolls sometimes had a coating of celluloid over the felt to make the doll easier to clean. The heads and hands were also sometimes made of celluloid.

The Raynal dolls carried several different identification marks. The tags read "Les Poupées Raynal/Narque Deposee/Made in France." The bottom of the shoe was often labeled "Raynal" and a labeled "Raynal" necklace also sometimes provided a marking.

The dolls look very much like Lenci dolls, especially the clothing. One distinguishing characteristic is their very wide eyebrows.

20.5" tall cloth molded face doll made by Edouard Raynal in France, circa 1920s. She is said to be a portrait doll of the actress Mary Pickford. She wears an original blue organdy dress with rows of ruffles on the skirt (much like one worn by Mary Pickford in one of her promotion photos). The hair on this doll is extremely coarse and feels much like the horse hair wigs found on mannequins. ($3,000-3,500). *Lazenby Collection.*

Germany

Käthe Kruse

Käthe Kruse initially began her doll enterprise when she made dolls for her own children. Her first unmarked dolls were shown at an "Art Dolls" German exhibition in 1909. This exposure resulted in an order from the FAO Schwarz toy store in New York City for 150 dolls. These dolls had the so-called frog hands (no attached thumb). The next doll, known as Doll I, had wide hips, six head seams, five leg seams, and an attached thumb. The cloth dolls had pressed cloth heads, oil painted faces, and the early dolls were 17" tall. These dolls in original clothing appear to be the most in demand by collectors although this design of doll was produced for many years into the 1930s.

According to Lynn Murray, writing in *Antique Doll Collector* in September 2002, the 13" Doll II (Schlenkerchen) had wire in his torso, arms, and legs, and an open-closed mouth. Doll III had wire armatures similar to the rare dollhouse dolls made by Käthe Kruse during World War I. Murray gives their number as IV. Both adults and children were produced.

The German firm also made two baby dolls. The smallest, an 18.5" tall doll, had closed eyes (Traumerchen # V) and the larger 23" tall baby was made with open eyes. The early dolls had painted hair and the later dolls were wigged, according to author Polly Judd. Sand was used as a weight in these dolls.

The 14" Doll VII was made using a small Doll I body with fewer seams but the early version with wide hips included the sewn-on thumb. The head was made from the Dumein model (see below). This doll was circa 1927. Later versions of the doll were slimmer. In 1930, the Doll I head was used on the doll.

Doll VIII (Dumein) was a larger doll, approximately 21" tall, which was first issued in 1929. This model was used for both boy and girl dolls. It was the first Kruse doll to have a swivel head.

The FAO Schwarz Christmas Catalog for 1930 listed several "Kaethe Kruse" dolls for sale, priced from $13.50 to $28.00 each. They came in 16", 18", and 20" sizes and included painted hair or the "New Kaethe Kruse Doll" with a specially constructed wig. All the dolls came in both boy and girl styles. The dolls with wigs were the new VIII style in the 20" size. They were more modern looking with respect to both clothing and hair styles. Both the boy and girl wore tan trench coats and red berets and represented older children than the molded hair dolls. The pictured boy and girl with molded hair were dressed in a bonnet and short dress (girl) and a romper type suit (boy).

After World War II, Käthe Kruse moved to West Germany and began re-establishing her business. It took several years, but the Kruse family members have been producing fine quality Kruse dolls for some time, many of which are based on the original designs.

The original Käthe Kruse dolls were stamped on the foot with the words "Käthe Kruse" and a number. The numbers do not indicate anything about the year they were produced.

14" tall Käthe Kruse Type X boy. His head is of molded fabric, painted with layers of oil paint. He was introduced in 1930. His hips are swivel jointed, and his arms are on tag-joints to give him a very floppy feel when he is held. ($1,500-2,000). *Photograph and doll courtesy of Nancy Lazenby.*

17" tall cloth Käthe Kruse Doll I model, made in Germany beginning in 1910. The body of these early dolls had wide hips, six head seams, five leg seams, and a thumb that was attached separately. The dolls were hand painted with oil paints and jointed at the shoulders and hips. They included black and brown eyeliner. This doll has been redressed. ($2,000-3,000). *Doll courtesy of Paula Reding and the Denver Doll Emporium.*

Margarete Steiff

Margarete Steiff began making dolls in Giengen, Wurtemberg, Germany in 1894, according to the Colemans writing in *The Collector's Encyclopedia of Dolls*.

The Steiff dolls can be easily recognized by the seam down the center of the face. A metal button in the ear was used as a company trademark, except on the very early products. The dolls were made of felt, plush, or velvet and had button eyes or glass eyes. Some had removable clothing and others had clothing that was a part of the body. Other characteristics include large hands and feet on some models. The Steiff dolls were jointed at the shoulders and hips and their heads turn. A lesser number were also jointed at the knees. The dolls came with painted hair or mohair wigs.

Johana Gast Anderton shows a *Playthings* picture of Steiff dolls from 1904 in her book, *The Collector's Encyclopedia of Cloth Dolls*. It shows the group of dolls that won a grand prize at the St. Louis World's Fair in 1904 and in Brussels in 1910. Two small child dolls, two adult male dolls, a horse, and smaller animals are pictured.

The firm made dolls representing real people (children, ethnic pairs, etc.) and more comic or caricature dolls, which included clowns, musicians, Golliwogs, gnomes, elves, and many more types.

According to Jan Foulke writing in *Doll Reader* in Aug.-Sept. 1986, Margarete Steiff died in 1909. Production continued until World War I , when few dolls were made. The firm resumed production after the war.

The Steiff company has re-issued some of its early dolls in limited editions for interested collectors.

This 10" tall Steiff boy with glass eyes was made by Margarete Steiff in Germany, circa mid-1920s. He is identified as an early doll because of the seams or darts coming in from the sides of his eyes. He wears his original costume (except for the shoes). His outfit is made entirely of cotton fabric, and consists of dark blue striped pants and a beige striped top with blue accents. His hat is of a medium brown color. ($1,000+). *Lazenby Collection.*

Italy

Lenci

The cloth (felt) Lenci dolls were made by a firm founded by Enrico and Elandi Scavini in Turin, Italy. These dolls were marketed beginning in 1920. The old dolls were reproduced beginning in 1978-79 for a number of years so that new collectors can also enjoy these newer "Art Dolls."

Early dolls from the 1920s and 1930s continue to be in demand by today's collectors, especially if they are all original. The dolls have felt swivel heads, are jointed at the shoulders and hips, and wear mohair wigs. The features were painted with oil paints and many of the eyes were painted looking to the side. The third and fourth fingers are sewn together and the dolls had applied felt ears.

Most of the early dolls were dressed in felt and organdy clothing and felt shoes. They were marked in a variety of ways including cloth or paper tags, a stamp on the bottom of a foot, or a tiny metal button. Since paper tags can easily disappear, many Lenci dolls are found unmarked.

Some of the first dolls were advertised in October 1920; Polly Judd pictures the ad in her book, *Cloth Dolls*. These dolls appear to resemble the Campbell Soup Kids and look very different from the Lenci dolls made a few years later. The ad states that the dolls were made of felt, are hand painted, and do not have a face seam.

Judd also pictures a Lenci ad from *Playthings* of January 1929. The dolls are from the firm's newly introduced "Sports Series." Pictured are a boxer, baseball player, tennis player, golfer, soccer player, and rower.

The Lenci dolls came in so many different sizes, costumes, and styles, a collector who wants to specialize in this area needs to research complete books on the subject as well as examining as many of the Lenci dolls as possible before making any purchases. Be aware that there are other felt Lenci "look alike" dolls that can be mistaken for a Lenci by both dealers and collectors who are not knowledgeable in the field.

Lenci dolls have been made in small sizes from 8"-9" tall (called Miniatures and Mascottes) and up to 40" tall in the long limbed Boudoir style dolls. Some celebrity dolls were also included in the Lenci line over the years. They include Rudolph Valentino and Mozart, both made in the 1920s.

The FAO Schwarz Christmas catalog for 1930 pictured three different styles of Lenci dolls. They included a 12" tall girl dressed in a felt dress, coat, and hat for $8.00, a 16" tall girl with long curls dressed in a peach organdy dress and large picture hat priced at $16.50, and a 17" tall boy doll dressed in a Scotch costume costing $26.50. Even in those days, the prices of the Lenci dolls were high. Effanbee Patsy and Horsman Dimples dolls cost only $10 each for their largest and fanciest models in the same catalog.

14" tall black Lenci court or street performer, circa 1920-1923. Typical of the earliest Lenci dolls, the clothing is part of the body construction, with flairs and ruffles added to simulate his jacket. His clothing is not removable. He is tagged with the tiny metal, Lenci inscribed, button attached to the back of his coat. ($1,000-1,300). *Lazenby Collection.*

22" tall Lenci clown dates from the earliest years of the Italian Lenci company. He does not appear in any catalog yet found. No other doll in the Lenci line has been found using the same face mold, so the sculpt is exclusive to this doll only. He dates from between 1920 and 1923. He is marked with a tiny metal button imbedded in the back of his ruffled collar that says "Lenci." ($3,500-4,000). *Photograph and doll courtesy of Nancy Lazenby.*

10" Lenci girl dating from 1923. She is tagged with the second type of Lenci metal button. It is much larger than the tiny one used on the earlier dolls. It is riveted to her underwear and says "Lenci" on it. This particular doll was an attempt by the Lenci company at producing an "economy model." The doll sold for $10 (wholesale), which was a great deal of money in 1923. Lenci dolls were meant to be art objects, and not particularly play dolls. ($400-500). *Lazenby Collection.*

14" tall Lenci 900 Series doll from 1929. These dolls all sit easily. Their floppy legs are on tab-joints making it easy to put them into a sitting position. She wears a felt trimmed organdy dress of pale pink, with lime green felt trim. Her hat is of organdy and felt with matching flowers. She is model number "H". Lenci's system used both letters and numbers to identify the different clothing styles. The system changed from year to year. ($750-1,000). *Lazenby Collection.*

The trunk sets first appeared in the 1930-31 catalogs. Some were tole-painted wooden chests while others were trunks. This Series 109 girl appears in the catalog but is identified only as "Series A." At 23" she is one of the largest Lenci children commonly found. She wears an intricately pieced turquoise coat with pink and white felt trim. Her hot pink dress is fairly simple. Her hat matches the coat. Also included is a green and white checked-pattern outfit consisting of a drop waist dress and separate jacket that fits over it. The outfit is completed with a hat and matching green shoes. ($3,000 with trunk). *Lazenby Collection. (Nancy Lazenby is currently writing a book on Lenci dolls.)*

The 8.5" to 9.5" Lenci Mascottes and Miniatures first appeared in the 1929 catalog and continued throughout the years as a staple in their line. This 9.5" doll is dressed in a royal blue tux, with a top hat and gloves pinned to his side. The tux is adorned in typical Lenci fashion with tiny roses at the waist and on the crown of the hat. ($400+). *Lazenby Collection.*

Advertising Dolls

The years from 1910 to 1930 provided a great many cloth advertising dolls for today's collector. Most of the dolls were to be assembled from printed fabric. They could be cut out, sewed, and stuffed at home.

Some of the most popular of the cloth advertising dolls from the period are those that were made to advertise Aunt Jemima Pancake Flour. According to Joleen Ashman Robison and Kay Sellers writing in their *Advertising Dolls* book, the first Aunt Jemima rag doll was issued in 1905, soon followed by four additional dolls in 1908. Dolls marked Davis Milling Co. date to before 1914, as the firm, located in St. Joseph, Missouri, became the Aunt Jemima Mills Co. after that time.

The four dolls issued by the firm were Aunt Jemima, Uncle Mose, Diana, and Wade. So many different sets of the dolls were made that without original advertising, it is hard to date the individual dolls. The firm was purchased by the Quaker Oats Co. in 1925, so when the dolls are marked with the company name, it helps to identify their period. At least seven Aunt Jemima dolls were made from 1905 until 1930.

The Cream of Wheat cloth advertising doll (called Rastus in a company ad in 1930) is also a prize doll for advertising doll collectors. Several different models of these dolls were marketed from approximately 1922 to 1952, according to Johana Anderton writing in her book, *Twentieth Century Dolls*. Many of the black dolls have photographic faces . In 1930, the 18" tall doll sold for 10 cents, according to a company ad from the November 1930 *Pictorial Review*. The ad stated: "New Big Rastus Doll/Get Yours Today! Rastus, the Cream of Wheat Chef in the form of a big 18" doll ... The Cream of Wheat Corp. Minneapolis, Minnesota."

The Kellogg Company offered a wide assortment of advertising dolls beginning in 1925. Their first series was based on the Goldilocks and Three Bears story. Four dolls representing the three bears and Goldilocks were offered. They ranged in size from 12" to 15" tall. In 1926, a different series of cloth Goldilocks and three bears became available to the public as advertising premiums. The Kellogg Fairyland Series of dolls was offered in 1928. These dolls were all 15" tall and included Red Riding Hood, Tom the Piper's Son, Mary Had a Little Lamb, and Little Bo Peep. The set cost 30 cents plus one box top from Kellogg's Corn Flakes.

Arkadelphia Milling Co. advertisement circa 1920 for their flour, featuring a cloth Dolly Dimple doll on each cambric flour sack. Most of the dolls were approximately 17" tall. When the sack was empty, the front and back of the doll could be washed, cut out, assembled, and stuffed to complete the doll. The advertisement comes from the Jaques and Tinsley Co. in Macon, Georgia and was featured in the newspaper.

Some of the many other cloth advertising dolls from this period include the following:

- 11" Dolly Dimple from the Arkadelphia Milling Co. in Arkadelphia, Arkansas. (Doll stamped on flour sack).

- 12.5" Miss Phoebe Primm from the Faultless Starch firm in Kansas City, Missouri circ 1914. Also printed Miss Lily White Doll. The Miss Phoebe Primm was reprinted in 1978.

- 26" Miss Korn Krisp of lithographed cloth, circa early 1900s. Advertising Korn Krisp Cereal.

- 13" Buster Brown doll advertising Buster Brown Shoes. Based on the character created in 1902 by Robert F. Outcault.

- 16" Ceresota Flour Boy circa 1912. He is marked "Ceresota Flour, the North Western Consolidated Milling Co." He was reproduced by Playthings By the Yard in 1973.

- Dy-o-la Dye Doll from the dye firm that was located in Burlington, Vermont, circa 1920.

An earlier advertising cloth doll was Miss Flaked Rice. She was pictured in an ad in *Cosmopolitan Magazine* in November 1900. This doll was a premium for Cook's Flaked Rice Co. in New York City. She could be purchased for only 10 cents and a product coupon. The doll was 24" tall and was to be cut out, sewn, and stuffed at home.

15" tall Aunt Jemima doll from the Aunt Jemima Mills Company in St. Joseph, Missouri. The doll's face appears to be a photographic image, unlike most of their dolls. The date for this doll is unknown but it is surely before 1920. ($100+). *Photograph and doll courtesy of Jean M. Grout*

Aunt Jemima Pancake Flour ad from the *Ladies Home Journal* of December 1924. The ad featured a set of four dolls that could be ordered from the Aunt Jemima Mills Co. in St. Joseph, Missouri for 25 cents and a box top from Aunt Jemima Pancake Flour or Buckwheat Flour. The dolls included Aunt Jemima, Uncle Mose, Little Diana, and Little Wade, and ranged in size from 12" to 15" tall. They came flat and were to be cut out, sewed, and stuffed at home.

15" tall Aunt Jemima doll similar to the one pictured in the *Ladies Home Journal* of 1924. She wears a different bandana on her head. "Aunt Jemima" is printed on her apron. ($75-100).

Uncle Mose cloth doll from the 1929 set of Aunt Jemima dolls. He is dressed in much finer clothing than the earlier dolls but the Aunt Jemima from the set is still attired in an apron and head scarf. Like all the other Aunt Jemima dolls, he came as flat fabric and was to be sewed and stuffed at home. He is marked "Uncle Mose" on the back on his collar. ($100). *Doll from the collection of Ellyn McCorkell.*

13" tall Goldilocks from the set of Kelloggs premium dolls dating from 1925. It cost 10 cents and one box top. The entire set included Goldilocks and the Three Bears. The set of four was priced at 30 cents and four box tops. "Kelloggs" is printed on the front and "Goldilocks" on the back. The set was made on flat pieces of fabric and needed to be sewed and stuffed at home. ($50 with soil).

Uncut Kelloggs advertising sheet of Little Bo Peep doll that measured 15" tall. She is from the 1928 Fairyland Series. Also included were Mary Had a Little Lamb, Red Riding Hood, and Tom the Piper's Son. Each was printed flat on fabric and made up as a 15" tall doll when sewed and stuffed at home. ($50).

Chapter Four

Composition Dolls

During the early 1900s, most of the dolls sold in the United States were imported from Germany. Although bisque dolls were the most popular, china, metal, and celluloid head dolls were also staples of the doll trade. By 1910, American toy companies were beginning to enter the doll manufacturing business by using a newly developed composition material. Although cloth dolls and papier mâché dolls had been produced in the United States for decades, it wasn't until the composition material was developed that many new companies began to manufacture dolls. The Aetna Doll and Toy Co. was one of the first. This firm was providing "Can't Break Em" composition dolls as early as 1909, according to an advertisement in *Playthings* magazine. The E. I. Horsman firm soon became the agent for the Aetna dolls. The first hit doll from this arrangement was the Billiken. Next came the popular Baby Bumps, then the Grace Drayton designed Campbell Kids in 1910.

Louis Amberg & Son (New York) began marketing composition dolls in 1910. The firm's first dolls were copied from German bisque models. By 1911, the firm was advertising Bobby Blake and Dolly Drake, who were designed by Grace Drayton. Louis Amberg & Son went on to manufacture many excellent composition dolls before they sold their line to Horsman, circa 1930.

Ideal Novelty & Toy Co. of Brooklyn, New York, also began manufacturing composition dolls at an early date. In 1910, the firm had their first hit with Mr. Hooligan, based on the comic strip "Happy Hooligan." The head was made in Germany, however. By 1912, the firm was marketing their own line of dolls, including Fan Kid and Naughty Marietta.

Fleischaker & Baum (Effanbee), of New York City, joined together in 1910 to produce composition dolls. By 1912, the firm was marketing composition dolls in both infant and child models. Included were memorable dolls like Johnny Tu-Face and Baby Dainty.

The Elektra Toy & Novelty Co., of New York City, was already making their well-known large composition carnival dolls in 1912. Besides these dolls stuffed with excelsior, the firm also produced all composition dolls.

The arrival of World War I and the resulting lack of dolls imported from Germany caused many more companies to enter the composition doll manufacturing field. Included were the Averill Manufacturing Co. in 1915, American Character Doll Co. in 1919, Natural Doll Co. circa 1915, Jessie McCutcheon Raleigh in 1916, as well as many others.

Firms that continued to produce composition dolls until the end of the era in 1950 included Horsman, Effanbee, Ideal, Arranbee, and American Character.

A few years after World War I ended, many of the new firms that had been established to make composition dolls found they could not adequately compete in the marketplace. This was mainly because German bisque dolls were once again being imported into the United States and these dolls were often priced lower than American made composition dolls.

Most of the American companies that produced top selling composition dolls throughout their era of popularity had their beginning in the 1910-12 era.

Louis Amberg & Son

Before this company was known as Louis Amberg & Son, Louis Amberg had been involved in the doll industry as an importer of dolls and toys beginning in the late 1870s. After working in partnership with several others from different companies, Amberg began manufacturing dolls for the firm known as Louis Amberg & Son in New York City circa 1911.

Some of the firm's early composition dolls were of better quality than those of their American competitors. Amberg hired Grace G. Wiederseim (later Drayton) to design Bobby Blake and Dolly Drake dolls for the company circa 1911. The dolls were based on her cartoon characters.

Several of Amberg's particularly collectible dolls from the period were made to represent celebrities. Included were Sis Hopkins (Rose Melville) in 1912, Charlie Chaplin also in 1912, John Bunny in 1914, and Baby Peggy made in 1923.

Other unusual dolls included the Victory Doll, the walking doll, and the AmKid doll, all circa 1918. These dolls were all produced to replace the German bisque head dolls that were no longer available because of World War I. The Victory doll had a composition wigged socket head with sleep eyes and a composition ball-jointed body. The walking doll came with either a composition socket head or shoulder head plus painted hair and eyes, or with a wig and sleep eyes. Both models had full composition arms, ball-jointed legs, and cloth bodies. The AmKid dolls were made with composition shoulder heads, leather bodies and legs, and full ball-jointed arms. These dolls had sleep eyes and wigs. (See "World War I Related Dolls" chapter for a photograph of the Victory doll.)

In 1914, Amberg marketed a doll with a bisque head called the Newborn Babe. It was designed by Jeno Juszko with a head made in Germany. The dolls was marked "LA&S/RA241 5/0 Germany." The doll had a bisque flange head, glass eyes, and an open mouth. The lower arms were composition and the body, legs, and upper arms were cloth. The doll looks very much like the Bye-Lo Baby doll, but was the first to be marketed. It was reissued in the 1920s to take advantage of the Bye-Lo popularity.

Several other dolls of particular interest were marketed by Amberg in the 1920s. The most popular is probably Mibs from 1921. She was designed by Hazel Drukker, author of the 1920 book *Phyllis May and Her Dollies*. Mibs was produced both in composition and in bisque. Sunny Orange Maid, circa 1924, was another unusual Amberg product. She was dressed in an orange dress and included an orange (fruit) "hat" on her head. Vanta Baby and Body Twist dolls, including Tiny Tots and "It," were also interesting dolls marketed by the Amberg firm as the decade of the 1920s came to an end.

In 1930, the composition line of dolls was sold to Horsman. That firm continued to issue several of the dolls including Vanta Baby, "It" (now named Peggy), and the smaller Tiny Tots dolls for several years.

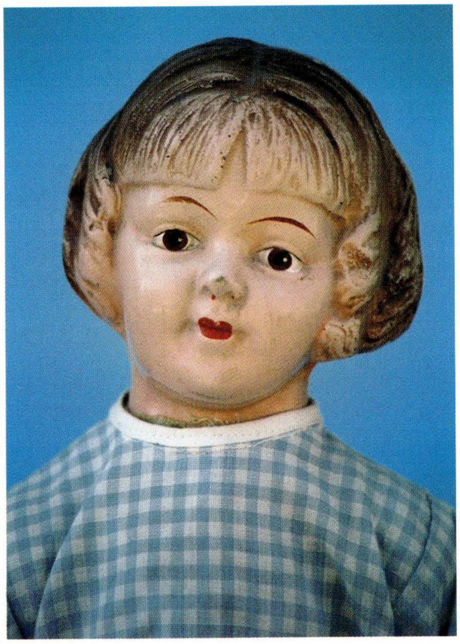

16" tall unidentified flange composition head doll, circa 1912-1915. The curls around her head look very much like those on the Curly Locks and Dorothy Dainty dolls made by Louis Amberg & Son in 1912. Both dolls were designed by Jeno Jusko. This doll has deeply molded and painted hair featuring bangs and curls, painted features, composition hands, and a cloth body, upper arms, and legs. Her body and upper arms are made of white cloth while her legs are pink. Her white cloth boots are made as part of her leg construction. She is jointed at the shoulders and hips with inside discs. If she is marked, it is on the back of her flared neck and is covered by her cloth body. The head is attached with lots of string. ($200).

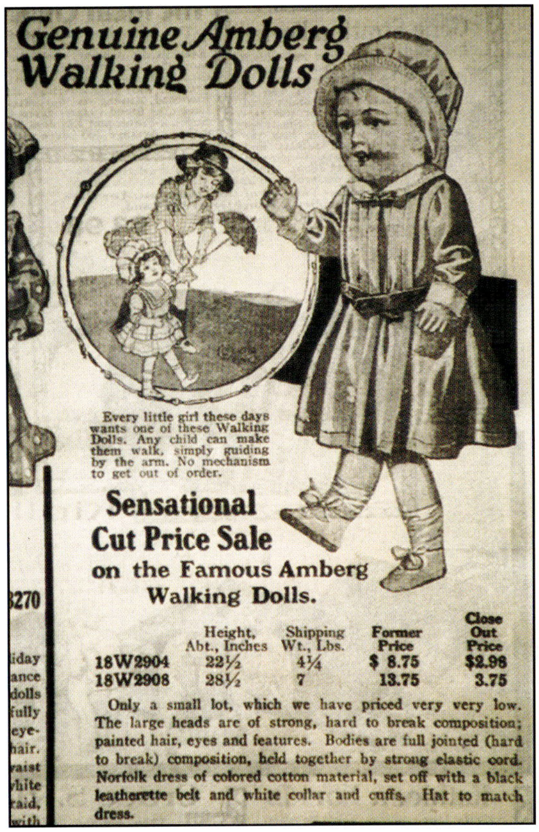

Right:
Advertisement for the Amberg Walking Dolls from a Sears Roebuck and Co. Sale Catalog of 1923. The dolls initially came on the market in 1919. They came in sizes of 22.5" and 28.5" tall and were being sold at the close-out prices of $2.98 and $3.75. The former prices were $8.75 and $13.75. The doll's heads were made of composition and they had painted eyes and features. According to the ad copy, the bodies were full jointed composition held together by elastic cord. The 1921 Sears catalog also listed a 19.5" tall Amberg Walking Doll on sale for $5.98. It had stuffed legs. The two larger models, as well as the smaller doll, came with mohair wigs and sleep eyes in this catalog. The dolls had full composition arms and the larger dolls had ball-jointed legs attached to a wooden platform that was secured to the bottom of the cloth bodies.

16" tall Mibs, a Phyllis May Dolly, from the Louis Amberg & Son firm, circa 1921-23. The doll was designed by Hazel Drukker and copyrighted in 1921. She has molded painted hair and painted features. The doll was made with a composition shoulder head and arms to above the elbows and unusual, specially designed composition legs with painted shoes and stockings. Some of the dolls came with regular shoes and socks. The body is cloth. Drukker was author of a book called *Phyllis May and Her Dollies*, published in 1920. A doll based on the Phyllis May character was also produced by Amberg in 1921. (See *Collector's Encyclopedia of Composition Dolls Volume II* by Ursula R. Mertz for a photograph of this doll.) The Mibs dolls came with a tag that read "Please Love Me I'm Mibs." The 16" size doll was sold by Sears in 1922 for $1.79. This doll wears what appears to be her original dress. ($800+). *Doll from the collection of Ellyn McCorkell.*

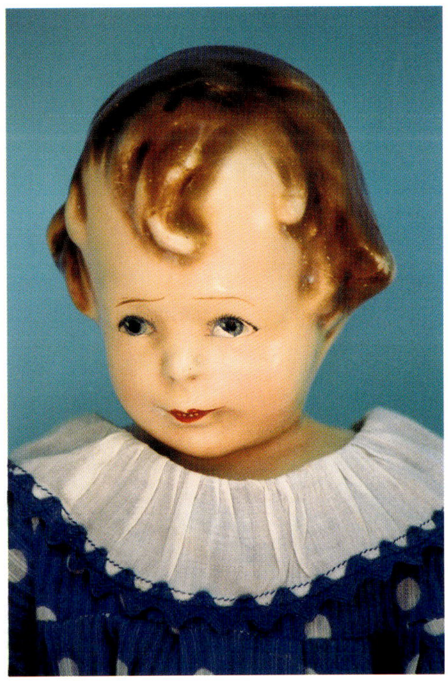

Close-up of the well-known Mibs' head. *McCorkell Collection.*

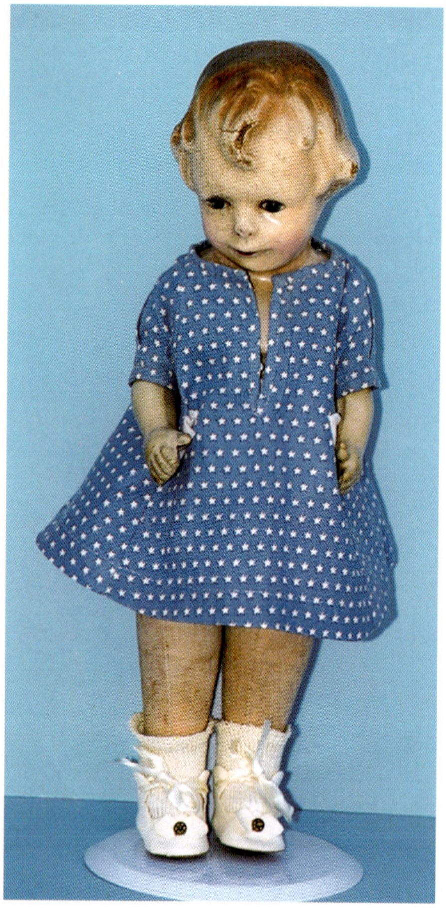

12.5" tall Mibs made by Louis Amberg & Son, circa early 1920s. This small version is seldom seen. The smaller Mibs has a composition shoulder head, molded painted hair, painted eyes with lots of painted eyelashes, and a painted closed mouth. She has full composition arms, cloth body, and cloth legs and feet. She is jointed at the shoulder and hips. She has been redressed. (With wear $300+).

8" tall, all composition Amberg Tiny Tots doll, circa 1928. These dolls came in both boy and girl designs. Both featured the famous patented body twist. The upper body, as well as the arms, can be moved. The legs and head are stationary. He has molded and painted shoes and socks. He is marked on his back "PAT.APPLIED FOR/LA&S © 1928." The body twist dolls were also made in a 13" size. (With wear $75+).

14" Amberg "It" doll as pictured in the 1929 ad. She is marked on the back "Amberg/Pat.Pend/ LA&S © 1928." The doll is all composition and has molded painted hair and painted features, with the special twist feature in her body. Her shoes and socks appear to be like those in the ad. Her vintage dress is similar to others worn by these dolls. ($450+). *Doll from the collection of Ellyn McCorkell.*

Amberg advertisement that appeared in *Child Life* magazine in November 1929. The Tiny Tots 8" size boy and girl dolls were priced at $2.00 for the pictured box set. The set included the boy and girl dolls along with outfits for each doll. The wardrobes featured coats and hats, two dresses, romper, and overalls. The 14" tall Amberg "It" doll was also pictured in the ad. She was an all composition doll with an unusual joint in her middle that allowed her to twist or bend. The doll sold for $3.00.

Close-up of "It" doll marketed by the Louis Amberg & Son firm in 1928 and 1929. *McCorkell Collection.*

21.5" tall Amberg composition shoulder head Vanta Baby doll, circa 1927-1929. This baby doll has molded painted hair, sleep eyes, an open mouth with two painted teeth, full composition arms and curved baby legs with a cloth body (including a small crier) and cloth upper legs. She is marked on the back of her shoulder plate "VANTA BABY/Amberg." She is all original, wearing a diaper, shirt, slip, and dress made in the Vanta manner, which used no pins or buttons for closures. Instead, each garment included ties. Her shoes, socks, and rattle are also original. The Sears catalog for 1927 listed three sizes of the dolls: 21.5", 19", and 14". The dolls pictured were dressed only in shoes, socks, diapers, and underwear. They ranged in price from $3.48 to $5.95. A dressed 12.5" Vanta Baby was shown in the 1928 Sears catalog. It included cotton stuffed legs and sold for $2.98. (With some wear $250).

Advertisement for Vanta Babies from the Sears catalog for 1930, with the dolls labeled as Horsman products. This was because Louis Amberg & Son sold their composition line to Horsman in 1930. Many of these dolls, including the Vanta babies, were then sold under the Horsman name. The babies in the ad were 19" tall and sold for $3.98 in Vanta's underclothes and $4.75 completely dressed. *Catalog from the collection of Marge Meisinger.*

The tag on the Vanta Dress reads "Vanta Baby/ Trademark Reg./An Amberg Doll.."

American Character Doll Co.

The American Character Doll Co. was just beginning business during the 1920s, having been founded in 1919 in New York City. The trade magazines *Playthings* and *Toys and Novelties* were both carrying the firm's advertising in 1920 and 1921. The ads pictured composition dolls both with and without wigs and moving eyes. By 1923, the firm was using "Petite" as a trademark, and many of the company's dolls can be identified by this mark.

The firm's popular baby dolls in the 1920s were called "Bottletot" and "Happy Tot." Each of the dolls included a hand molded to hold a bottle. A "Tottle Tot," issued circa 1930, did not feature a bottle in its hand.

American Character also issued many different models of "Ma-Ma" dolls during the 1920s. As with all other companies, the early models were short and wide while later dolls featured slim bodies. All of these dolls featured criers.

The company surely produced a popular doll in 1927 when they came out with "Lucky Aviation Kid" to take advantage of Lindberg's successful transatlantic flight. The Colemans picture the June 1927 *Playthings* ad in their *Collector's Encyclopedia of Dolls, Vol. II.*

Some of the firm's most unique dolls of the period were advertised in *Playthings* in 1929. These included the Puggy Doll, which appears to have been based on the drawings of Charles Twelvetrees. The newly authorized American Character Campbell Kids dolls, which were no longer being produced by the E.I. Horsman firm, were also from this period.

Sally, a new Petite all composition doll, was advertised in 1930. She had molded hair and sleep eyes, and was made to compete with Effanbee's popular Patsy doll line.

See *Dolls & Accessories of the 1930s & 1940s* (Schiffer Publishing, 2002) for information and photographs of additional American Character composition dolls.

13" tall American Character composition Wonder Baby, circa 1926. She has molded painted hair, sleep eyes, and an open mouth. The baby has a composition shoulder head, full composition arms, cloth body, and curved legs. Her right hand is also curved so she can hold a bottle. The doll is all original except for the socks, plus she is missing her bottle. This doll is also known as Bottletot. ($150).

American Character Doll. Co. ad for Bottletot that appeared in *Playthings* in November 1926. The copy reads "Bottletot/ A Petite Baby/infringes on no other doll patents. It is an original baby…The bottle is removable from the hand…The doll is COMPLETE with or without the bottle. Our own design patents are pending." Sears catalog offered an 18" tall Bottletot doll priced at $4.48 in 1926.

12" tall American Character all composition Puggy doll, circa 1929-30. He is marked on the back "A Petite Doll." He has molded painted hair, painted eyes, and a closed mouth. He is jointed at the shoulders and hips and is wearing his original clothing. A similarly dressed Puggy was advertised in *Playthings* magazine in December 1929. He resembles artist Charles Twelvetrees' drawings of the period. (With craze and separation lines at the back of the head, $250)

American Character ad for two of their dolls that appeared in *Junior Home Companion* in December 1928. The pictured dolls include a "Ma-Ma" doll that came in various sizes priced from $4.50 to $7.95 each and a Happy Tot Baby doll. The baby dolls were priced from $2.50 up. The trademark for the American character dolls was "Petite." The ad copy states that "All Petite dolls have head, legs, and arms of unbreakable composition." The Happy Tot dolls also had the right hand molded to hold a bottle. The dolls had full composition shoulder plates to below the arms. Sears sold the "Happy Tot" doll in 1927 for $1.98 for the 12.5" size and $3.98 for the 16" size.

American Character ad from *Playthings* in May 1930. The ad featured the firm's new Sally doll (to compete with Effanbee Patsy). The all composition doll had molded painted hair and sleep eyes. Sears carried the doll in their 1930 catalog priced at $3.49 for a 19" tall model. The Baby dolls pictured were called Toddle Tots. These dolls had composition flange heads, composition arms and legs, and cloth bodies. Their hair was molded and they had sleep eyes and open mouths with teeth.

Averill Mfg. Corp. (Mme. Hendren)

The Averill Mfg. Corp was founded circa 1915 by Paul and Georgene Averill (husband and wife) in New York City. The firm was advertising in *Playthings* as early as 1916. Many of their first composition dolls were dressed as Indians or Dutch Boys or girls in costumes made of felt.

Many of the Averill dolls were sold using the "Madame Hendren Dolls" trade name. The firm is credited with inventing the popular "Ma-Ma" dolls that featured a crier by 1920.

Other famous dolls from the era included the Dolly Record dolls advertised in 1922. These dolls play cylinder records and were similar to Effanbee's Mae Star dolls. The composition whistler dolls marketed by the firm are also very collectible. They were patented in 1926 and included the black Rufus and dolls dressed as cowboys or sailors.

Averill was also responsible for a celebrity doll from the period that represented Betty Bronson, a young actress of the time. This topsy-turvy doll was advertised in *Playthings* magazine in July 1926.

Besides designing their own dolls, the company also produced Madame Hendren dolls using the designs of other artists. Grace Cory's Little Brother and Little Sister were special dolls of this type that date from 1927. Grace Drayton, another artist from the period, also designed dolls for the firm. Dolly Dingle and Chocolate Drop dolls were produced from her designs.

The cartoon character Snookums was marketed as a Madame Hendren doll circa 1927. According to Ursula R. Mertz writing in her book *Collectors Encyclopedia of American Composition Dolls Vol. I*, these newer dolls were based on the 1927 Universal movie *The Newlyweds and Their Baby* and represented the child who played the Snookums baby in the film.

Besides designing composition dolls for her own firm, Georgene Averill also designed dolls made by other companies during this time period. Included were the Pinky Winky dolls (see "Celluloid Dolls" chapter), the bisque head Bonnie Babe doll (see "Bisque Baby Dolls" chapter) and the Sunny Boy and Sunny Girl dolls, which featured celluloid heads made in Germany. All of these dolls are also very collectible.

Non-Breakable Character Dolls

COWBOY DOLL AND PREPAREDNESS KID

These two dolls are new this season and are making a great hit. Their popularity is instantaneous. The pictures do not nearly show how attractive they are. They both have unbreakable heads and hands with durable bodies and legs. Their clothes are very attractive, too. One is dressed like a real cowboy with a pistol and leather holster on his belt. The other is dressed like a soldier in uniform. They are the famous Madame Herndon character dolls and you will not be disappointed with either of them. They measure 9½ inches tall.

Price $1.00. Sent postpaid.
Either doll given free for two new yearly subscriptions.

THE PETERKIN TWINS

These dolls are just as cute as their picture shows them. There are no more clever dollies than these. They have "Can't Break 'em" heads, arms and legs. They are attractively dressed: the girl, Betsy Peterkin, in a pretty bonnet with strings tied under her fat little chin and a pretty serviceable dress and socks, and Tommy Peterkin wears a cunning, soft hat, rompers and socks. Both have very attractive round chubby faces and pink and white complexions. You will love them at first sight. They are good-sized dolls and extremely serviceable.

These dolls were designed by Mrs. Dayton who also designed the famous Campbell Kids. We believe they will be more popular than any dolls we have ever offered.

Price $1.50. Sent postpaid.
Either given free for three new yearly subscriptions.

THE DUTCH DOLLIES

These dollies are dressed like little Dutch children and are unusually attractive. Their pictures do not show one-half how attractive they are. The boy has broad Dutch trowsers of a bright color. The little girl wears just the style of dress that the little girls in Holland wear and they both have quaint Dutch hats. They are the famous Madame Herndon character dolls with non-breakable heads and hands.

Price $1.00 each. Sent postpaid.
Given free for two new yearly subscriptions.

Address all orders to LITTLE FOLKS MAGAZINE, SALEM, MASSACHUSETTS

Advertisement circa 1916 picturing several Averill Mfg. Co. dolls, including a pair of Dutch dolls, a cowboy (featuring a gun and holster), and what appears to be a soldier. The dolls had composition heads and hands. The cowboy and soldier dolls measured 9.5" tall each and could be purchased for $1.00 each. No size is listed for the Dutch dolls, but since they are also priced at $1.00 each, it is likely that they were approximately 9" tall.

13" tall "Soldier" doll, circa 1915-1916, thought to have been marketed by the Averill Mfg. Co. The same doll was also sold by the firm dressed in felt costumes representing a cowboy and a Dutch girl. The firm may have purchased the dolls from another company, dressed the dolls, and sold them under the Averill name. This example has a composition flange head with composition hands. His hair is molded and his features are painted. His made-on-shoes are black cloth. The body, legs, and upper arms are also cloth. His outfit is made of felt and appears to be original. The arms and legs are jointed with outside disks. The doll also still has his original belt and (not pictured) holster. The hatband matches his belt. One of his arms has come loose and hangs below his sleeve. ($75 in worn condition).

18" tall composition shoulder head doll made by the Averill Mfg. Co., circa 1920. She is marked on the back of her shoulder plate "A.M. Co." The doll has a mohair wig, sleep eyes, closed mouth, composition arms nearly to the shoulder, and swing composition and cloth legs. Her body, upper arms, and legs are cloth. She is a "Ma-Ma" doll and has a crier in her back. Her clothing is original (except for replaced ribbon). She wears a one piece underwear set and a matching dress and bonnet. Her shoes have been replaced. The Averill firm has been credited with inventing the "Ma-Ma" doll with a crier in her body. ($150).

Madame Hendren "Ma-Ma" advertisement from the 1923 Sears Fall & Winter catalog. The pictured doll had painted eyes, a wig of mohair, composition head, arms, and legs, and a cloth body with a crier. The dolls sold for $2.39 for the 14.5" size and $2.98 for the 16" size.

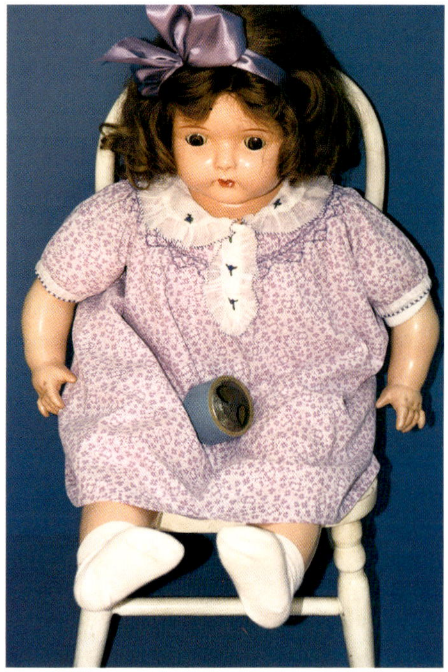

22" tall Madame Hendren Dolly Record doll, first produced in 1922. She has a composition shoulder head, arms, and lower legs, with a cloth body. The doll has sleep eyes, an open mouth with two teeth, and a human hair wig. "Madame Hendren" was stamped on the back of the dolls' cloth bodies. A mechanism was placed in the back of the cloth bodies allowing cylinders to be inserted into the "phonograph" to play nursery rhymes. The mechanism appears to be the same as that used in the Effanbee Mae Starr dolls. This doll has been redressed. ($550). *Doll from the collection of Ellyn McCorkell.*

The Sears 1923 catalog also featured the Madame Hendren (trade name for Averill) Rock-a-Bye Baby doll. The doll cried as she was being rocked to sleep and said "Ma-Ma" when she was picked up. She came in sizes of 16", 18", and 20". Prices listed were $6.75, $7.98, and $9.98.

Mechanism used in the Dolly Record doll, with a cylinder inserted that plays the song "One-two." The cylinder is marked Averill Manufacturing Co. The cylinder in the doll's lap plays "Mother Hubbard." *McCorkell Collection.*

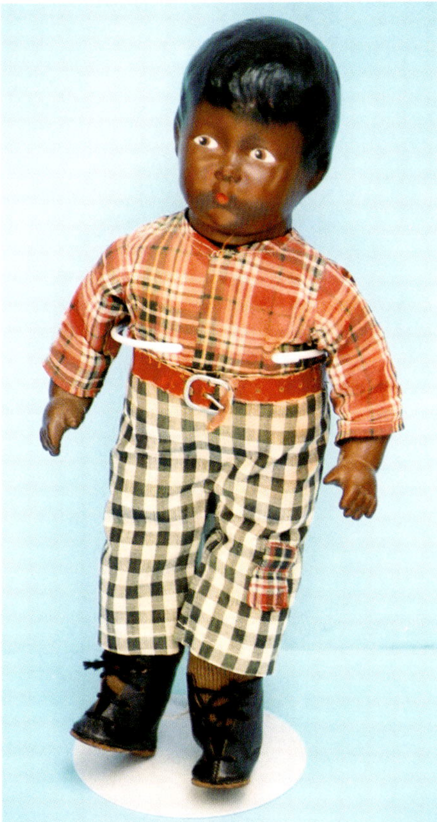

Madame Hendren's 13.5" tall black whistling Rufus, patented in 1926. He has a composition flange head, painted hair and features, with a molded puckered mouth in a whistling position. The doll has full composition arms and a cloth body and legs. His legs contain bellows and steel springs. He whistles when the legs are pressed down on a table. His outfit appears original. ($300). *Doll from the collection of Ellyn McCorkell.*

Madame Hendren advertisement from the *Ladies Home Journal* of November 1927, advertising Sunny Girl (there was also a Sunny Boy). This was an unusual doll for this period in America because she included a celluloid shoulder head made in Germany. (See "Celluloid Dolls" chapter.) She had an American made body including composition arms and lower legs and a cloth body and upper legs. Her cloth body included a crier. The dolls came in sizes from 14" to 22" tall.

14" tall Little Brother designed by Grace Corry for Madame Hendren, circa 1927. He has painted molded hair and features, a composition shoulder head, composition arms and legs, and a cloth body and upper legs. He is marked "©/By/Grace Corry." He has been dressed as a pirate and was called Captain Kidd in one of the company ads. ($400+). *Doll from the collection of Ellyn McCorkell.*

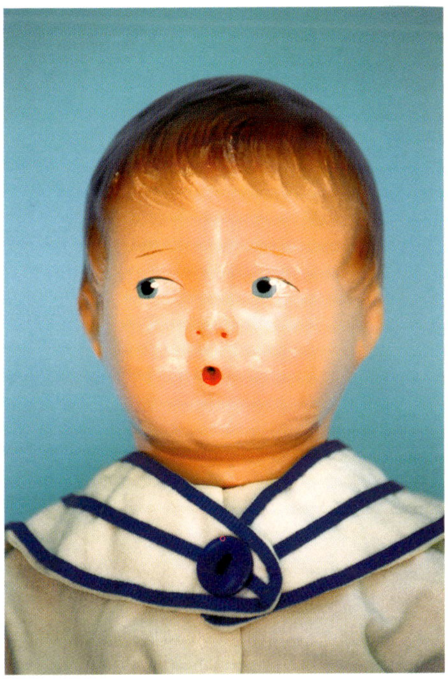

13" Hendren Whistler showing a close-up of the molding of the lips. This one is dressed as a sailor. *McCorkell Collection.*

Right:
14" tall Little Sister, in this case dressed in a Valentine costume, designed by Grace Corry for Madame Hendren. She has a mohair wig, painted features, a composition shoulder head, composition arms and legs, and a cloth body and upper legs. She is marked "©/By/Grace Corry." She is all original, wearing a red felt Valentine shaped dress and matching red felt shoes. The original tag reads "Be My Valentine." It is likely that she was a special Valentine promotion. A 1929 *Playthings* ad features Little Sister holding an Easter basket, evidently another special doll geared to a holiday. ($450+). *Doll courtesy of the Denver Doll Emporium and Paula Reding.*

This advertisement for the Madame Hendren LYF-LYK Baby Brite doll appeared in the *Ladies Home Journal* of December 1928. The baby dolls had composition flange heads and limbs with cloth bodies, molded hair, sleep eyes, and open mouths with two teeth. They came in seven sizes. The coupon at the bottom of the ad entitled the sender to receive a free golden good luck coin. For only ten cents a cloth cut-out Dolly Dingle doll would be included, which could be stuffed and made into a cuddly doll.

Effanbee – Fleischaker & Baum

The Fleischaker & Baum firm was founded in New York City by Bernard E. Fleischaker and Hugo Baum in 1910. Using the initials of the two founders' last names, the company's trademark became "Effanbee." According to the Colemans writing in their book, *The Collector's Encyclopedia of Dolls*, the firm registered the trademark in the United States in 1918 although the Effanbee name first appeared in advertising in 1913.

Effanbee composition dolls are among the most valued by today's collectors. Even the firm's early dolls are much sought after. Particularly desirable are the various models of Baby Grumpy, Johnny Jones, Coquette, Pouting Bess, Billy Boy, Betty Bounce, Aunt Dinah, Uncle Mose, Johnny-Tu-Face, and their whistling dolls. These dolls were made with composition heads and hands along with cloth bodies and limbs. The faces and hair were painted.

By 1915, some of the dolls (particularly babies) had wigs. A more modern Effanbee wigged Mary Jane doll was introduced in 1917. She had a composition socket head, ball-jointed composition body, and sleep eyes. (See chapter on "World War I Related Dolls.") The firm's dolls continued to become more modern through the late teens, with many equipped with wigs and sleep eyes. By 1922, "Ma-Ma" dolls were being added to the doll line, and the slogan "They Walk, They Talk, They Sleep" had become an accurate description of most of the firm's dolls.

Popular dolls of the era included Baby Dainty, Harmonica Joe, new versions of the earlier Grumpy dolls, Pat-O-Pat (claps hands when chest pushed), Rosemary and other "Ma-Ma" dolls, and the Mae Starr phonograph doll issued in 1928.

The most famous Effanbee doll of the 1920s was the baby doll called Bubbles, first introduced circa 1924-25. There were many versions of this doll, including both a curved leg baby model and a straight leg toddler model. The dolls were also made in a variety of sizes.

As the 1920s drew to a close, the Effanbee firm began marketing a new, all composition 14" doll called Patsy in 1928. This model and the other dolls in her line became Effanbee's biggest sellers for years to come. See *Dolls and Accessories of the 1930s & 1940s* (Schiffer Publishing, 2002) for more information and photographs of these dolls.

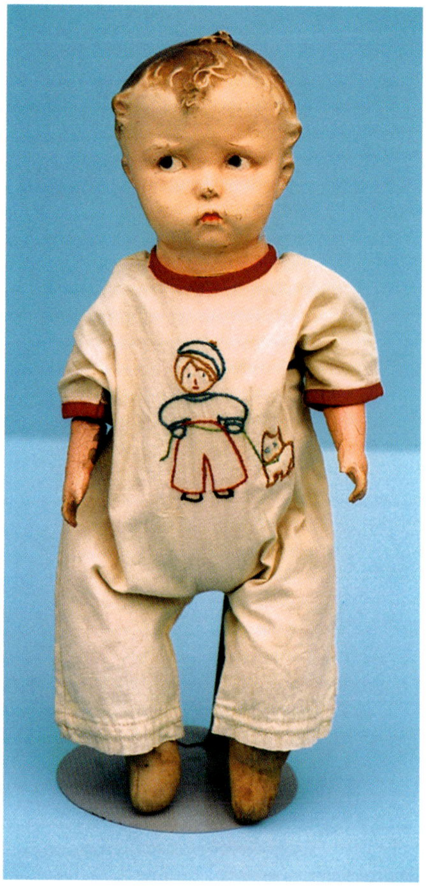

Right:
14" Effanbee Baby Grumpy with a flange composition head, circa 1915. The firm's 1915 catalog stated that the doll came in two sizes, priced at 65 cents and $1.25 each, in an assortment of costumes. The sizes were not listed. The doll has molded painted hair, intaglio painted eyes, closed mouth, lower composition gauntlet hands and lower arms, and a cloth body, upper arms, and bent cloth legs. He is marked "174" on the back of his neck and is jointed at the hips and shoulders with inside discs. He has been redressed in a romper of the period. ($200+).

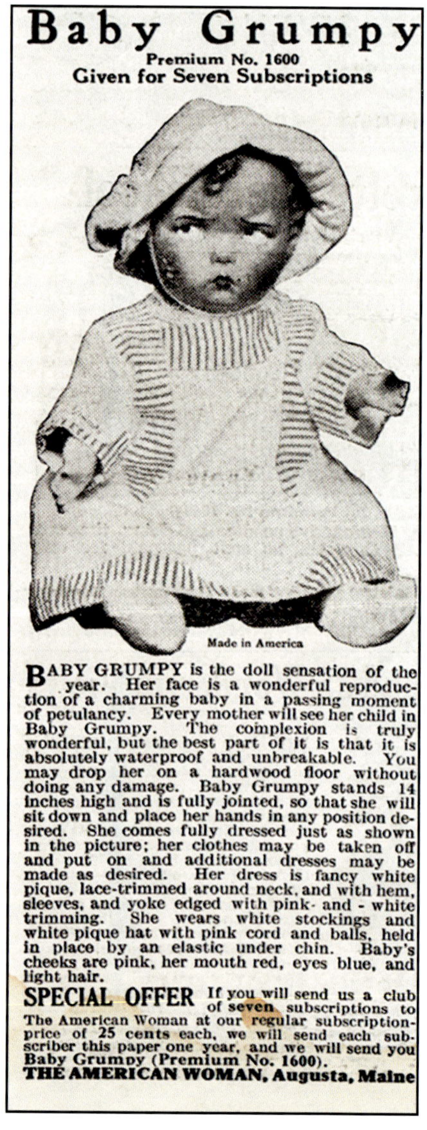

The Baby Grumpy was offered as a premium in the *American Woman* magazine in October 1915. The copy read "Baby Grumpy is the doll sensation of the year." This doll was 14" tall and could be obtained by sending in seven new subscriptions to the magazine at 25 cents each.

11" tall black doll marked on the back of the neck "F+B," circa 1916-1918. She has a composition flange head and hands, and a cloth body, arms, legs, and feet. She is jointed at the shoulders and hips with inside discs. Her legs are made of striped material to represent stockings and she has black fabric sewed on shoes. There is a loop for a hair bow on her molded hair. She has painted features with a closed mouth. This doll has the look of a Grace Drayton design. The modified F & B Grumpy doll used for the black Aunt Dinah doll was also made with these same striped legs. ($200-250). *Doll from the collection of Ellyn McCorkell.*

21.5" tall Effanbee shoulder head baby doll, circa 1916-1919. She has painted blue eyes, closed mouth, and her original brown mohair wig. She is marked "Effanbee" in script on the back of her shoulder head. The doll has lower composition arms with cloth upper arms attached with inside discs. Her body and curved legs are cloth, also mounted with inside discs. The cloth body is in mint condition with no accumulated dirt or dust. Her bootie-socks appear to be original as does her underwear. Her dress, slip and bonnet may also be original but they are made of nice cotton instead of the cheaper coarser material usually used for doll clothing of the period. She has some paint loss under one eyes. ($150+).

Effanbee (Fleishaker & Baum) ad from *Toys and Novelties* in September 1919. Two baby dolls and three little girl dolls are pictured. All look similar to the German bisque head dolls, although the Effanbee line was made of composition. The doll on the left appears to be Mary Jane (ball-jointed body). See "World War I Related Dolls" chapter for a photograph of this doll.

Left:
14" Effanbee Whistling Jim doll, circa 1916. He has a composition head with molded hair and painted features including intaglio eyes and an open whistling mouth. His lower arms and hands are also composition. The upper arms, body, and legs are cloth (ground cork stuffed). His black cloth shoes are made as part of the legs. He originally had a whistling mechanism in the body. He is in his original clothing that once carried a cloth tag reading "EFFANBEE/Whistling Jim/Trademark" on the front of his bib. Otherwise he was unmarked. ($300-400).

Right:
Effanbee black model of Harmonica Joe with a composition shoulder head, circa 1924. He is marked "Effanbee" in script on the back of his shoulder. He has molded painted hair, painted eyes, full composition arms, cloth body, and legs. His rubber mouthpiece and original harmonica are missing. He has a rubber ball in his stomach, which was to be pressed to make the harmonica on his mouth play. He is jointed at the shoulders and hips with inside discs. His overalls are original but his shirt, shoes, and socks have been replaced. ($75-100).

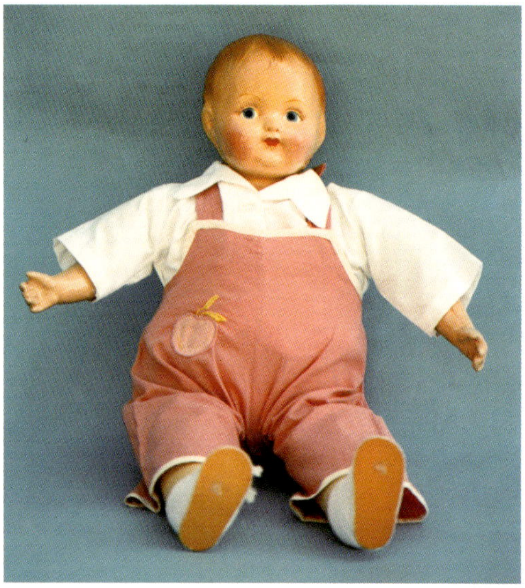

16" tall composition shoulder head doll, circa 1920s, marked Effanbee in script on the back of his shoulder plate. He has composition lower arms and a cloth body and legs. There are no shoulder joints. The hair is molded and the features are painted (includes some repainting). He has been redressed. He originally wore rompers and a hat. ($75).

20" tall Effanbee composition shoulder head "Ma-Ma" doll, circa early to mid 1920s. She is marked "Effanbee" in script on the back of her shoulder plate. She has sleep eyes, her original mohair wig over molded hair, closed mouth, composition lower arms and composition feet and legs past her knees. The body, upper arms, and upper legs are cloth. A crier is mounted in her back. She has no shoulder joints and has swing legs. She has been redressed. ($100+).

12" tall, all composition Black Grumpykins doll, circa mid 1920s. She has painted molded hair and painted features. The doll's body is similar to the Bubbles dolls from the same period. She has a composition head molded with a larger upper body piece that extends almost to her waist. Her head does not turn. The shoulder is marked in the back in a circle "EFFANBEE/Dolls/Talk Walk Sleep." The lower body is cloth containing a crier and she has full composition arms and legs. She has been redressed in a commercially made dress. ($200).

Effanbee "Ma-Ma" dolls were advertised in the Sears catalog for 1923. The doll was 29" tall and sold for $8.95. The dolls had a "Ma-Ma" crier.

14.5" tall Baby Dainty composition shoulder head doll made by Effanbee, circa early 1920s. She has molded painted hair, painted features, full composition arms, cloth body, and cloth legs. The back of her shoulder head is marked in script "Effanbee." She is jointed at the arms with inside disc joints and her cloth legs are sewn on. She wears her original one piece underwear and dress. ($100-125).

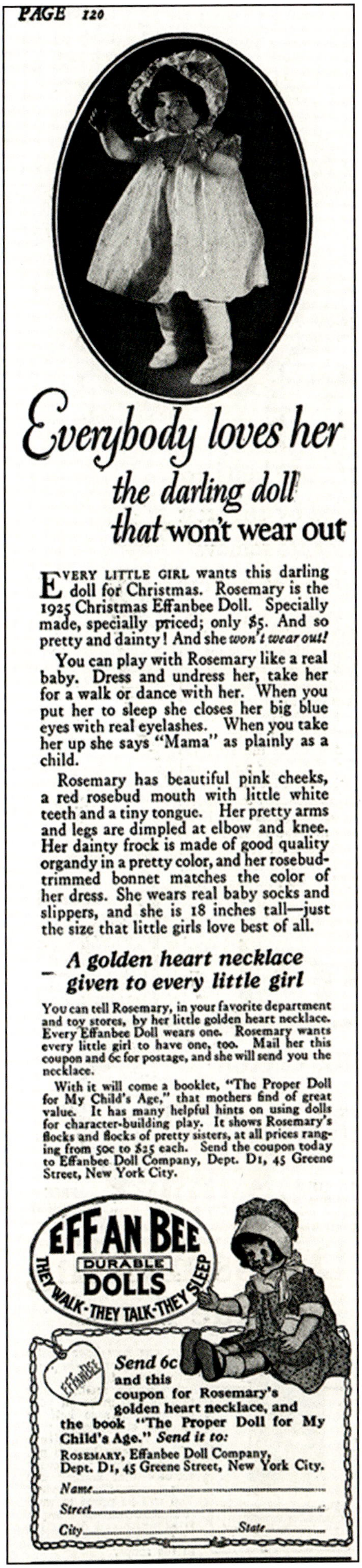

Left:
Effanbee "Rosemary" ad from the *Junior Home Companion* in October 1925. The doll was 18" tall and cost $5.00. She was described as having sleep eyes, open mouth with teeth, and was dressed in an organdy dress and bonnet. Each of the dolls of this type came with a heart shaped necklace marked "Effanbee."

17" tall composition shoulder head doll, circa 1925, marked "Effanbee/Dolls/Walk Talk Sleep" in a raised circle on her shoulder plate. She has sleep eyes, open mouth and teeth, and a mohair wig. The doll has composition arms and lower legs with a cloth body. She is jointed at the shoulders and hips. She wears her original organdie dress and bonnet as well as her "Effanbee" labeled heart shaped locket. ($200+). *Doll from the collection of Jan Hershey.*

Right:
30" tall Effanbee Mae Starr composition shoulder head doll, circa 1928. She is marked on the shoulder plate "Mae Starr." She has sleep eyes, open mouth with teeth, real hair wig, full composition arms, and lower composition legs. The mechanism to play the records is contained in her cloth body (back). She came with two of her original records. She is wearing her original dress, slip, and bloomers. ($600+).

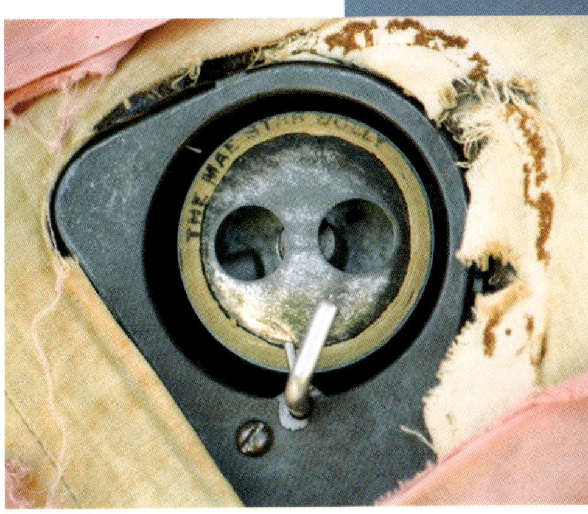

One of the doll's records marked "The Mae Star Dolly" is pictured inside the phonograph mechanism. A crank in her side can be used to make the record play. These seem to be the same records that were used with the earlier Averill Mfg. Co. Dolly Record doll.

Two different versions of the Effanbee Bubbles dolls, circa mid 1920s. The larger doll is 16" tall and is marked on the back of her shoulder plate in an oval EFFANBEE/Dolls/Walk.Talk.Sleep" and under the oval "Made in USA." She has a composition head and body to under her arms, painted molded hair, sleep eyes, open mouth, composition full arms, curved baby legs, and a cloth body with a crier in the back. She is jointed at the shoulders and hips. The smaller doll is a 13" tall toddler version of Bubbles. The back of her shoulder head is marked "EFFANBEE/Bubbles/Copy R. 1924/Made in U.S.A." The difference in the two dolls is in their leg styles. Both dolls have cloth bodies with criers and upper bodies made of composition, like that of the Grumpy doll. Both have been redressed in commercially made clothing of the period. A 14" Bubbles sold for \$2.59 in 1927. (Redressed \$150 each).

Effanbee Bubbles ad that appeared in *Playthings* magazine in November 1926. The dolls had been on the market since 1924, but were not heavily advertised until late 1925 and 1926. This ad stated that the dolls came in five sizes.

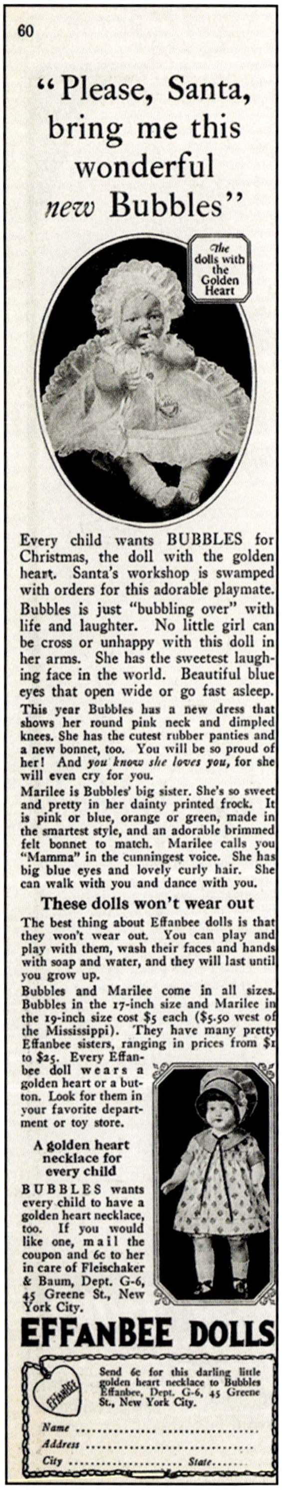

Left:
Effanbee ad from the *Junior Home Companion* in December 1928. It features the Bubbles doll in the 17" size priced at \$5.00. Other sizes were also available. The Marilee "Ma-Ma" doll was also pictured. She came in a 19" size also priced for \$5.00. Both dolls came with the Effanbee heart necklace, which could be ordered for only six cents. The copy in the ad stated that Effanbee dolls could be purchased priced from \$1.00 to \$25.00 each.

Right:
19" tall Effanbee composition shoulder head "Ma-Ma" doll, circa mid to late 1920s. She is marked on her back in script "Effanbee." She has tin sleep eyes, open mouth with teeth, and full composition arms. Her swing legs are made mostly of composition with cloth upper legs that connect to the cloth body. The body is slimmer than those from earlier years. It also contains a crier. Molded unpainted hair is concealed under her real hair wig. She has been redressed in commercial clothing of the period. (With paint flecking on arms \$75-100).

E.I. Horsman Co.

The E.I. Horsman Co. was founded by its namesake in New York City, circa 1865. According to Don Jensen writing in *Collector's Guide to Horsman Dolls*, the firm sold bisque head dolls imported from Germany as early as the mid 1870s.

The first doll success Horsman had under its own brand name was the line of cloth Babyland Rag Dolls beginning in 1893. These dolls were sold by Horsman until circa 1914. (See "Cloth Dolls" chapter.) By 1909, Horsman was contracting with the Aetna Doll & Toy Co. to make composition dolls for the Horsman firm.

One of the dolls made with this arrangement was the popular Billiken figure. It had a composition head and a plush Teddy Bear type body. He came in 12" and 15" sizes. The next successful composition Horsman doll was Baby Bumps, introduced in 1910. The doll had a composition flange head and a cloth or plush body and limbs. He looked very much like Kämmer & Reinhardt's model #100 bisque head baby.

Also in 1910, the Horsman firm received a license to produce dolls based on Grace Drayton's drawings of the Campbell Kids (used for advertising Campbell's Soup). The early dolls had composition flange heads and cloth bodies and limbs.

According to author Don Jensen, Helen Trowbridge became a famous doll designer for the Horsman firm in the early years. Some of her designs included Baby Butterfly, a Japanese baby from 1914, Candy Kid (1910), Little Sunshine (1913), and Peterkin (circa 1910).

Comic characters of the day were also translated into doll form by the Horsman firm. The Gene Carr Kids, Little Nemo, Little Mary Mix-Up, and Ella Cinders were all dolls based on comic strips of the day. (see "Comic and Celebrity Dolls" chapter for more information.)

For a short period, circa 1915-1916, Horsman also marketed pull toys with dolls attached. These types of toys were made by several companies under different trade names, including Coaster Kid and Irish Mail Kids.

In addition to selling composition dolls, Horsman also played a part in the Fulper Pottery Co. of Flemington, New Jersey's production of bisque head dolls during the World War I years. The arrangement was short lived and ended in 1919. Also in 1919, Horsman and Aetna merged their companies and the products eventually carried the better known E.I. Horsman logo in advertising.

Like most of the doll companies in the United States, Horsman began producing various "Ma-Ma" dolls during the 1920s. Dolly Rosebud was the trade name for their most popular model.

In 1921, a Jackie Coogan doll was marketed by the Horsman firm to tie-in to the Charlie Chaplin film called *The Kid*. The young Jackie Coogan played an important role in the new movie. This doll was designed by Helen Trowbridge and came in sizes of 13.5" and 15.5" tall. (See "Comic and Celebrity Dolls" chapter.)

Several of the Horsman baby dolls from the 1920s were unique and now include a variety of desirable dolls for today's collector. Probably the hardest to find is Baby Horsman, designed by Edith Hitchcock and first offered in 1923. This doll had interesting molded hair with a curl on his forehead. The baby included a composition head, composition hands, and a cloth body and limbs. The collectible Tynie Baby dolls were copyrighted in 1924. They came with a composition flange head, composition full arms, and a cloth body and legs. The babies had molded hair, sleep eyes, and a closed mouth. The most popular of the 1920s baby dolls was Dimples, first issued in 1927. A variety of different models and sizes were issued into the 1930s. The earliest dolls had composition flange heads, full composition arms, and bent composition legs to above the knee, plus a cloth body and upper legs. The dolls had molded hair, tin sleep eyes, and an open mouth with teeth. Fewer of the dolls were made with straight legs, and later Dimples dolls featured swing arms and legs with the upper parts made of cloth.

Other memorable Horsman dolls from the 1920s included HEbee-SHEbees, based on the characters of Charles Twelvetrees circa 1926 (see "Comic and Celebrity Dolls" chapter) and newly designed Peterkins Dolls from 1929.

The Horsman firm continued to produce dolls for decades to come but, for the most part, the company's "glory days" were over by the 1930s. Most of their new dolls could not compete with top of the line dolls marketed by high end companies like Madame Alexander and Effanbee. Horsman found its niche by producing cheaper lines of dolls that were sold in mail order catalogs and at "ten-cent" stores instead of in department stores.

12" tall Billiken "Teddy Doll" with a composition head and a plush body, circa 1910. He is jointed at the shoulders and hips and has molded painted hair and painted features. He still has his original tag reading "The Billikin Company/Chicago." The Billiken was very popular at the time and many different products were manufactured featuring a Billiken image. ($350+). *Doll from the collection of Ellyn McCorkell.*

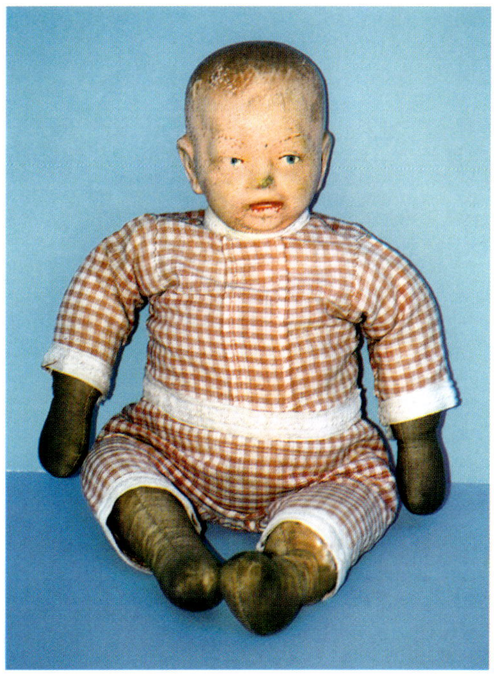

12" tall unmarked Horsman Baby Bumps with a composition flange head, circa 1910. He has painted slightly molded hair, painted eyes, closed mouth, and a cloth body with curved legs. He is jointed at the shoulders and hips with inside discs. He wears his original romper suit. (With wear $85-100).

Advertisement for Horsman dolls from the 1914 Montgomery Ward catalog. The composition dolls ranged in size from 10.5" to 15" tall and were priced from 48 cents to 98 cents each. Included were Little Sunshine, Little Baby Blossom, Billy Boy, Sunshine Baby, and Sunbonnet Sue. *Catalog from the collection of Marge Meisinger.*

Other Horsman dolls from the 1914 Montgomery Ward catalog included these two baby dolls. The smaller doll was 13.5" tall and sold for 98 cents. The larger Baby Blossom was 17" tall and priced at $1.95. *Meisinger Collection.*

12" tall Horsman Candy Kid, designed by Helen Fox Trowbridge in 1910. He has a composition flange head, painted hair and features, and a cloth cork stuffed body, arms, and legs. His cloth shoes were made as part of his body. He wears his original rompers. ($200+). *Doll from the collection of Ellyn McCorkell.*

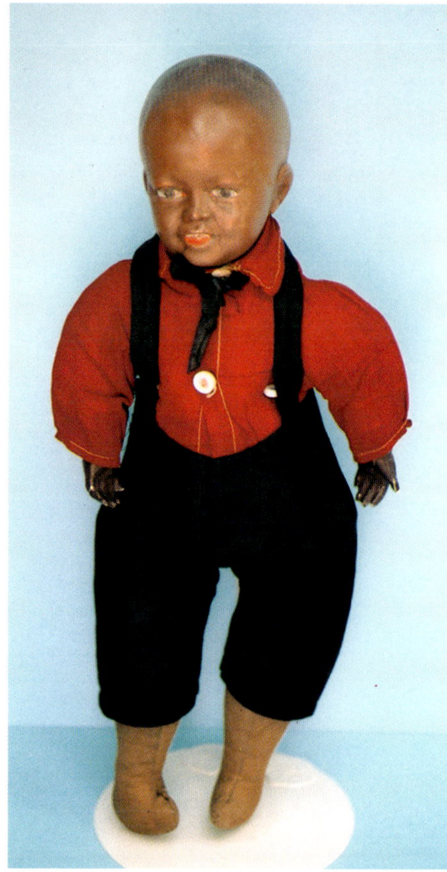

14" tall Horsman Black Cotton Joe doll, circa 1914. He has a composition flange head and hands, painted hair and features, and an open-closed mouth with painted teeth. The doll has a cloth body, upper arms and legs and is wearing what appears to be his original clothing. ($250+) *McCorkell Collection.*

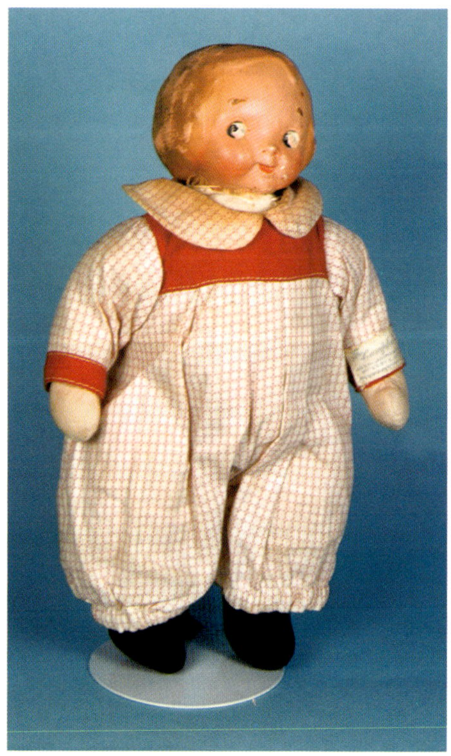

10" tall Horsman Campbell Kid designed from the drawings of Grace Drayton, circa 1910. These popular dolls remained in the Horsman line for many years. In 1928, the American Character Doll Co. secured rights to the dolls. The pictured early example of the Horsman boy doll includes a composition flange head with painted and molded hair and painted features. The rest of the doll is cloth. The black cloth shoes are made as part of the body. The tagged outfit is original and identifies the doll as a Campbell Kid. ($300+). *Doll from the collection of Marge Meisinger. Photograph by Carol Stover..*

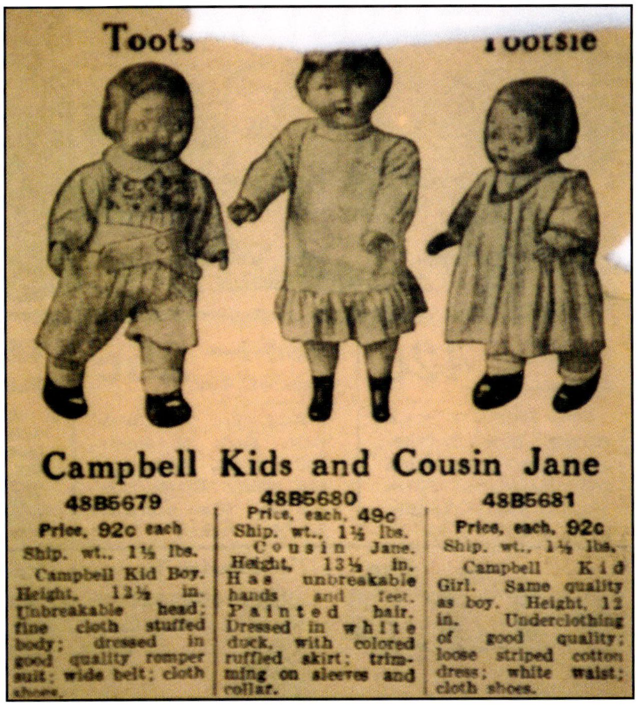

Advertisement for Horsman Campbell Kids' dolls, which appeared in the Montgomery Ward catalog in 1916. Both the boy and girl dolls were available in 12.5" sizes priced at 92 cents each. The boy appears to be wearing the outfit that was patented by Horsman and included fabric designed with drawings of the Campbell Kids used as decoration.

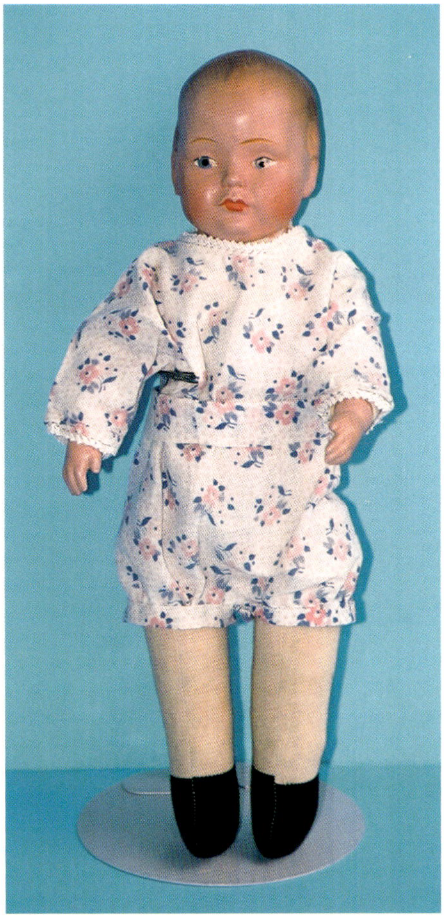

This doll appears to have a head like the doll pictured in the Jensen Horsman book as one of the Irish Mail Kids. He identifies the head as a Horsman "Our Baby" head. The unmarked pictured doll is 12" tall and has a composition flange head with gauntlet composition hands. He has painted hair and eyes and a closed mouth. The cloth body, arms, and legs are in excellent condition with no dust or dirt having accumulated through the many years since he was manufactured. Unfortunately his original clothing is missing.($75).

13" tall possibly Horsman doll with composition flange head circa 1917. The black doll had molded hair, painted features, gauntlet composition hands, and a cloth body, upper arms, and legs with "made on" black cloth shoes. She is jointed at the shoulders and hips with outside discs. The back of her head is marked "HC(I)Q/1917" with the smaller letter I within the C. Don Jensen writing in his book *Collector's Guide to Horsman Dolls 1865-1950* indicates that the mark might stand for "Horsman Co. Inc. Quality" but the mark has not been positively identified. She has been redressed. ($100+).

Advertisement for the Coaster Kid from the Montgomery Ward catalog in 1916. This is one of the pull toys from the period that is attributed to Horsman. The others are the Irish Mail Kids and the Velocipede Kids. Effanbee also marketed similar toys called Cycle Boy and Dixie Flyer. Various models of dolls were attached to the toys. The one pictured was priced at 99 cents. It measured 11" high by 11.5" long.

Advertisement from the *Little Folks* magazine, circa 1917. The Horsman Peterkin twins are pictured. The copy states that the dolls were designed by Grace Drayton. The price was listed as $1.50 but it probably means each. The dolls were 11" tall and included composition heads.

19" tall Horsman composition shoulder head doll, circa early 1920s. The doll has sleep eyes, open mouth with teeth, original mohair wig, composition lower arms and hands, and a cloth body, upper arms, and swinging cloth legs. An early cry box is in the back of the doll. He is marked on the shoulder plate "E.I.H. © A.D.Co." The A.D. Co. may represent the Aetna Doll Co., which had merged with Horsman in 1919. He wears his original tagged Romper suit and hat. The tag reads "Horsman/Doll/Voice Patented/By B.E. Lloyd." ($100+).

Horsman "Ma-Ma" doll advertisement in the Sears 1923 Fall and Winter Catalog. She was made with a composition shoulder head, sleep eyes, and a mohair wig. The doll had composition arms and the larger dolls had composition legs. The bodies on all the dolls were cloth as were the dolls upper arms and legs. Each came with a crier box in the cloth body. The dolls ranged in size from 14" to 18" tall. They were priced from $3.48 to $5.65 each.

19" tall Horsman "Ma-Ma" doll with composition shoulder head, circa 1924. She has sleep eyes with real eyelashes, open mouth with teeth, and her original brown mohair wig. The doll was made with full composition arms, composition feet and legs to above the knee, and cloth upper legs. Her cloth body contains a crier box. She is marked on the back of her shoulder plate "EI © H Co/E. I. Horsman/INC." She is all original except for her hair bow. She is wearing a pink dress, slip, and bloomers in a one piece undergarment, and her original large shoes and socks. All of the "Ma-Ma" dolls from this period had very large feet. ($200+).

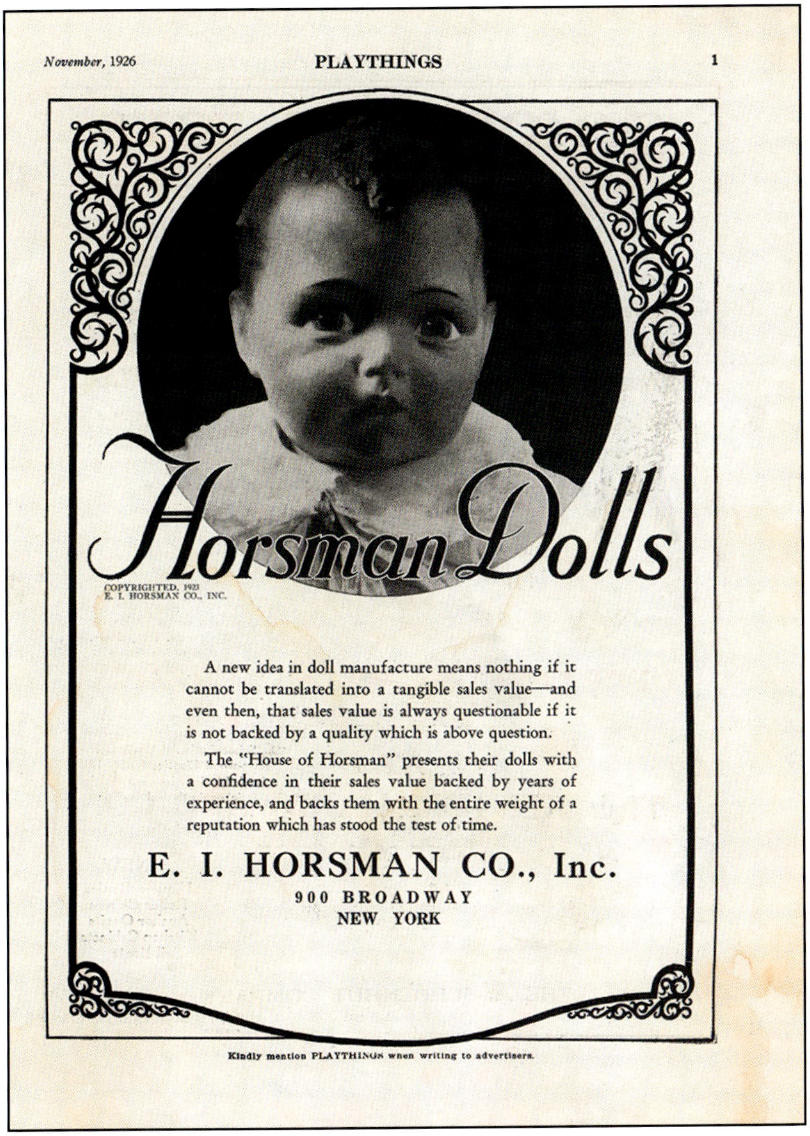

Ad from *Playthings* issue of November 1926. Although the doll pictured is not identified, it is the doll known as Baby Horsman and under the picture is "Copyrighted 1923/E.I. Horsman Co., Inc." The doll was made with a composition flange head (shoulder head on the small models) and hands or lower arms on the smaller models, plus a cloth body and limbs. The doll pictured had very distinct molded painted hair, painted eyes, and a closed mouth. It was also produced with sleep eyes. The doll came in a number of sizes including 14", 20", and 24".

This Horsman advertisement appeared in the *Ladies Home Journal* in December 1928. It pictured the Horsman Baby Dimples as well as the firm's Dolly Rosebud dolls. The Dimples doll was first introduced in 1927. It came in sizes from 16" to 22" tall. An all composition straight legged Dimples was also made. Later the Baby Dimples doll was produced with swing legs. Dolly Rosebud represented the doll industry's newer design of the "Ma-Ma" dolls. The newer dolls had slimmed down bodies and smaller feet. The pictured doll was probably 16" tall.

18" tall Horsman composition flange head Dimples, circa 1927. She is marked "© E.I.H. CO.INC" on the back of her neck. She has molded painted hair, sleep eyes, open mouth with teeth, full composition arms and composition legs to above the knee. The body and upper legs are cloth and a crier is located in the body. She wears her original one piece underwear and dress. The dress label reads "Horsman/Doll/MF'D in U.S.A." The shoes are replacements. ($175 with some wear).

13.5" tall Horsman Peterkin dolls, circa 1929. The dolls include composition flange heads and limbs with molded and painted hair and painted features. The bodies are cloth. They are marked "E I H © INC." The girl is wearing her original tagged costume. The "boy" has been transformed from another Peterkin Indian girl with the addition of new pants trimmed with the original fringe from the girl's skirt. ($600 pair). *Dolls from the collection of Ellyn McCorkell.*

Close-up of the tag on the Peterkin girl. It reads "Trademark/Peterkin." *McCorkell Collection.*

18" tall Horsman baby doll thought to be from the Gold Medal line, circa 1931. She has a composition shoulder head, molded painted hair, sleep eyes, and an open mouth with teeth. Her lower arms and swing legs are composition and the upper limbs and body are cloth. The body includes a crier. She is all original. The dolls originally came with a gold medal that pictured a baby on one side and a Swastika on the other (which once represented good luck). ($200+). *Doll from the collection of Ellyn McCorkell.*

Ideal Novelty & Toy Co.

The Ideal Novelty Company was founded by Morris Michtom and Aaron Cone in 1907 in Brooklyn, New York. Michtom had already been producing Teddy Bears since 1903. The firm's first doll was the Yellow Kid, marketed in 1907 and based on the character in Richard Outcault's popular comic strip of the same name. In 1910, a new doll called "Mr. Hooligan" (from the comic strip "Happy Hooligan") was issued by the firm. An Ideal advertisement in *Playthings* in January 1910 pictured the Mr. Hooligan doll with an Ideal unbreakable head and also advertised an unpictured "Snookums, the Newlywed" doll with a head made of Ideal unbreakable composition. (See "Comic and Celebrity Dolls" chapter.)

By 1911, the Ideal Co. was producing a whole line of "unbreakable" composition dolls. Popular dolls included one based on Ty Cobb, the well-known baseball player, and one called Baby Mine.

In 1912, according to Judith Izen writing in *Collectors Guide to Ideal Dolls*, the partnership broke up and the name of the firm became Ideal Novelty and Toy Co. New innovations continued in the company when Ideal received a patent for moving eyes that closed individually in 1915.

Ideal continued to add interesting dolls to their line in the teens. Included were Farmer Boy in 1914, with a puckered mouth (for whistling) and a molded hat; Naughty Marietta, circa 1912, whose interesting feature was a depression around the head to be fitted with a ribbon; and Liberty Boy, circa 1917. (See "World War I Related Dolls" chapter.)

Ideal dolls based on famous products also became popular in the teens. Uneeda Biscuit Kid was advertised in *Playthings* in June 1915. This doll was based on a character used by the National Biscuit Co. in their advertising. Also issued was a doll that advertised Zu-Zu Ginger Snaps. She was dressed in a yellow clown suit with brown stars and held a small box advertising the cookies.

Other important Ideal dolls of the period included Soozie Smiles (a two-faced baby), circa 1923; Flossie Flirt, circa 1925; Smiles (baby), 1927; Buster Brown, circa 1929; and Tickle Toes, beginning in 1929 and continuing into the early 1930s.

The Ideal Novelty & Toy Co. continued to be successful in the doll and toy manufacturing arena for many more decades. See *Dolls & Accessories of the 1930s and 1940s* for more photographs and information about Ideal's composition dolls, including their most famous doll of all, based on the child star of the 1930s—Shirley Temple.

16" tall unmarked flange composition head doll, circa 1915. The doll appears to be an Ideal product with its molded boots and brown socks. The composition head has a molded space around her head to be fitted with a hair ribbon. She has painted features, gauntlet composition lower arms and hands, and a cloth body, upper arms, and legs. She is jointed at the shoulders and hips with inside discs. A similar head was made with sleeping eyes. She has been redressed in an old dress from the period. ($125-150).

15" tall Ideal Farmer Boy, circa 1914. He has a composition flange head with painted features and molded and painted hair. He also has a molded red cap on his head. His mouth is molded in a whistling position. The doll has composition arms to above the elbow and composition legs to above the knee. He is barefoot, has a cloth body and upper limbs, and is jointed at the shoulders and hips with inside discs. He is wearing his original clothing. ($300+). *Doll and photograph from the collection of Susie's Museum of Childhood at Bluebird Farm in Carrollton, Ohio.*

Ideal's Uneeda Biscuit Kid as advertised in the Montgomery Ward catalog in 1916. He was 15.5" tall and was copied from the "little fellow that advertises Uneeda Biscuits." He was priced at 99 cents.

15" tall Uneeda Biscuit Kid doll made by Ideal, circa 1915. The doll advertised a National Biscuit Co. product. He has a composition flange head, painted hair and features, composition lower arms, and a cloth body, upper arms, and legs. He has tall molded boots on his feet. He is jointed at the shoulders only with outside discs. The doll is all original and wears his yellow raincoat and hat with a romper suit under his coat. He carries his original Uneeda Biscuit "box" under his arm. It is a paper covered wood block that contains the same information as the original large packages, including the price of 5 cents for a package of Uneeda Biscuits. The original cracker boxes to go with the dolls are very hard to find. (Doll $350+ with box).

12" tall Ideal composition flange head doll, circa 1915. She is marked on the back of her neck in a diamond "Ideal" and outside the diamond "U.S. of A." The doll has molded hair, sleep eyes, open-closed mouth with teeth, a cloth body and upper legs, and molded brown boots. She has full composition arms that are attached outside with discs on one side and metal on the other. The legs are attached in the same manner. The long black stockings are original but she has been redressed ($75-100).

Left:
13" tall Ideal flange composition head boy doll, circa 1917, in original clothes. Although he is unmarked, he has the sleeping eyes patented by Ideal and wears the molded brown boots used with many of the early Ideal dolls. He has molded and painted hair, closed mouth, gauntlet composition hands, and a cloth body, upper arms, and legs. One of his hands has been replaced. He is jointed at the shoulders and hips with outside metal pieces. His original two-piece sailor costume was made very cheaply. ($200).

Right:
Ideal advertisement from *Toys and Novelties* magazine of September 1919, advertising the sleeping eye dolls made by the firm. The ad indicated that the company also sold the heads only, to be used on other doll bodies. *Ad from the collection of Marge Meisinger.*

15" tall Ideal composition shoulder head doll, circa 1918. She is marked "Ideal" in a diamond and "U.S.of A." She has molded hair, tin sleep eyes, an open-closed mouth, and composition hands. Her body, upper arms, and legs are cloth. She is jointed at the shoulders and hips with inside discs. The doll has been redressed. ($75-100).

Ideal "Walking and Talking" doll advertised in the Sears 1923 Fall and Winter catalog. The composition head doll was 18" tall, had sleep eyes, said "Ma-Ma," and featured a human hair wig. She was jointed at the shoulders and hips and had a cloth body. The doll sold for $3.98.

13.5" tall Ideal all original doll, circa 1926. He has a composition flange head with sleep eyes, closed mouth, and a mohair wig. The lower arms are composition while his upper arms, body, and swing legs are cloth. The cloth body includes a crier. He is wearing his original romper, hat, shoes, and socks. The tag reads "Ideal" (in a diamond) and "I Can Talk/I Can Walk/I Can Sleep." ($125 +).

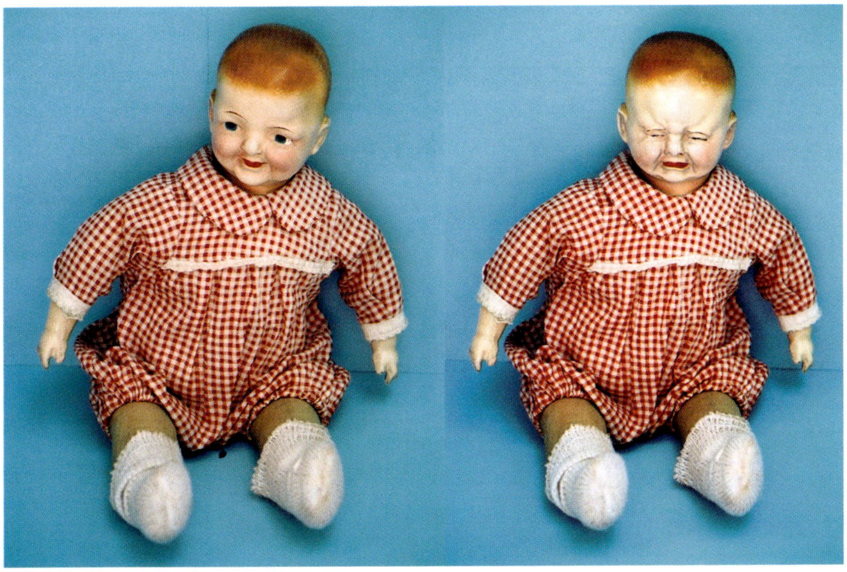

Ideal 16" tall Soozie Smiles doll, circa 1923. He has a composition flange head with two faces, painted features, composition arms, and cloth body and legs. One face is smiling and the other face is crying. He also has a crier box inside and is wearing his original rompers but is missing his hat. ($300+). *Doll from the collection of Ellyn McCorkell.*

This advertisement for Soozie Smiles appeared in the Sears 1923 Fall and Winter catalog. He appears to be wearing the same style romper as the previous Soozie. The doll was supposed to say "Ma-Ma" when picked up and cried when being put down. The doll sold for $2.35.

Advertisement in *Needlecraft* magazine of August 1926. The pictured Ideal doll could be obtained by selling four yearly *Needlecraft* subscriptions for 50 cents each. For two extra subscriptions, the additional dress and rompers would also be included. The doll (called Rosy-Cheeks in the ad) looks very much like the Ideal doll in the previous picture. She has the same characteristics and includes the identical tag as the one that came with the Romper boy.

Another similar Ideal doll was advertised as a free gift in *Farm Life* magazine of June 1927. She was called "Dolly May." The coupon included in the ad had to be filled out and returned to *Farm Life* to learn how the doll could be obtained "without spending a cent." It was probably given in return for a certain number of new subscriptions to the magazine.

15" tall Ideal "Smiles" doll, circa 1927-29. This near MIB doll is marked in a diamond "IDEAL." The doll has a composition flange head, rubber arms, cloth body and legs, flirty sleep eyes, and an open mouth with two painted teeth. She is all original including her pacifier. (Near MIB $300+). *Doll from the collection of Ellyn McCorkell.*

Original box and insert paper that came with the Ideal Smiles doll. A 14" tall Smiles with rubber arms and legs sold for $2.98 in the Sears 1928 catalog. *McCorkell Collection.*

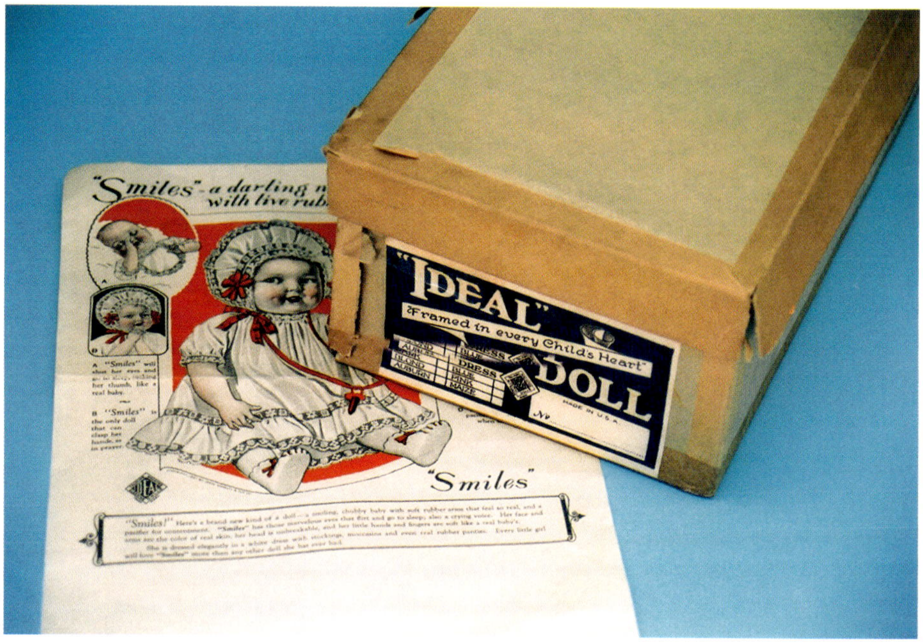

Left:
17" tall Ideal Peter Pan doll, circa 1929. He has a composition flange head, lower composition hands, cloth body, legs, and arms, molded hair, and sleep eyes. He is marked "Ideal" in a diamond. He is missing his original clothing, which included a hat, slippers, vest and collar. This doll wears a replaced vest and hat. ($150-175). *McCorkell Collection.*

Right:
Ideal Flossie Flirt doll, circa 1930. She is marked on the back of her composition flange head "Ideal" in a diamond and "U S of A" outside the diamond. Flossie has a composition flange head, a mohair wig, flirty sleep eyes, open mouth with teeth, nearly full rubber arms and legs, and a cloth body, upper arms, and legs. She is a "Ma-Ma" doll and has a crier located in her back. Her clothing, shoes, and socks are all original. She wears bloomers attached to a slip and a voile dress. ($125-150).

Miscellaneous Composition Dolls

Arranbee Doll Co.

The Arranbee Doll Co. was founded in March 1919 by William Rothstein in New York, according to authors Suzanne L. DeMillar and Dennis F. Brevik writing in their book, *Arranbee Dolls*.

At first the new firm concentrated on selling imported German bisque dolls. The biggest success the company had in those years was My Dream Baby, a bisque head baby doll produced by the German Armand Marseille firm. (See section on "Bisque Head Baby Dolls" for a photograph of this doll.)

By the later 1920s, the firm had begun the manufacture of composition dolls, including baby dolls and little girl "Ma-Ma" dolls. They had even more success with their composition dolls produced during the 1930s and 1940s, as well as with hard plastic dolls in the 1950s. The company was sold to Vogue Dolls circa 1958.

Berwick Doll Co.

The Berwick Doll Co. of New York City held a patent for the "Famlee Doll," which was issued in 1923. According to the Colemans writing in their book, *The Collector's Encyclopedia of Dolls*, the first patent had been issued to David Wiener in 1921, at which time the dolls were produced by the Change-O-Doll Co. A 16" tall composition doll with a cloth body was used as the basic doll. A variety of heads were produced, which could be screwed onto the basic doll's body like a light bulb, thereby changing the doll into various characters. Included were a clown, baby, boy, Indian, and other models. Extra clothes for these heads were also provided in boxed sets, which came with as few as one and as many as eighteen extra heads.

14" tall Arranbee composition flange head baby doll, circa late 1920s. He is marked on the back of his neck "ARRANBEE." The baby has molded painted hair, sleep eyes, open mouth with two painted teeth, hands of rubber (somewhat melted), cloth body, upper arms, and curved cloth legs. He is jointed at the shoulders and hips with inside joints, and his cloth body includes a crier. His clothes may be original but he is missing his hat. ($50-75 with rubber damage).

16" tall Famlee Doll from the Berwick Doll Co. based in New York. The doll has a composition head that was screwed into her body similar to the way a light bulb is screwed into a socket. She has a mohair wig, painted features, composition arms and legs, and a cloth body. Other heads in this set include a clown and a baby. Clothes for these two heads were also part of the original boxed set. ($300+)

Information on the box includes the following information. The "Famlee Doll" was patented on Apr. 12, 1921 and the Berwick Doll Co. held a patent issued in 1923. The outside of the box reads "Copyright 1923 Berwick Doll Co." The largest number of heads listed in a set is eighteen. The original price of the set was $6.00.

Fiberoid Doll Products Co.

The Fiberoid Doll Products Co. was very active in the manufacturing of composition dolls during the 1920s but apparently many of their labeled dolls were finished and marketed by other companies. Pam and Polly Judd, writing in their book *Composition Dolls Volume II*, picture a tagged Alexander "My Betty" Doll marked "Fiberoid" on the back of her shoulder plate. It is likely that the firm also sold doll parts, including heads, to other firms as well. The Judds mention that Fiberoid was a member of the "Doll Parts Manufacturing Assoc."

Several of the Famlee doll sets were advertised in a special Sears Fall catalog in 1926. Included were a set featuring only one extra head priced at $2.98, a set with two extra heads for $4.98, and a set with four extra heads for $8.39. Each set included clothing for the extra heads. Heads included clowns, Indians, babies, little girls, a navel officer, and more. *Catalog from the collection of Marge Meisinger.*

16" tall Fiberoid composition shoulder head twin dolls, circa early 1920s. The backs of the shoulder heads have holes at each corner to allow the head to be sewn onto a doll body. The cloth bodies are very worn although they appear to be factory made, with mitten type hands and cloth feet. The hips are very wide. The heads have molded painted hair, painted features with blue eyes, and closed mouths. They are wearing their homemade dresses, probably made by the original owners. ($50 pair).

21" tall composition Fiberoid shoulder head "Ma-Ma" doll, circa late 1920s. She is marked "Fiberoid" on the back of her shoulder plate. She has a real hair wig, sleep eyes, open mouth with teeth, and full composition arms and legs to above the knees. Her body and upper legs are cloth. The doll wears her original green organdy lace trimmed dress and one piece slip and bloomer underwear. Her shoes and socks have been replaced. ($100).

Eugene Goldberger

The Eugene Goldberger firm, located in Brooklyn, New York, was in business beginning around 1917 and remained active in doll production into the 1930s. Most of the dolls were not marked but the firm produced many different doll designs during the period. Included were carnival dolls, baby dolls, and "Ma-Ma" dolls. Beginning circa 1923 the EEGEE trademark was used for the firm's dolls.

13" tall Goldberger composition shoulder head doll marked "Goldberger" on the back, circa 1920. The doll has molded hair and painted features, lower composition arms, and a cloth body, upper arms, and legs. The ivory cloth shoes are made as part of the legs. She has been redressed in vintage clothing. ($50-75).

Natural Doll Co.

The Natural Doll Co. was established in New York City, circa 1915. The firm made inexpensive dolls into the 1930s. According to Ursula R. Mertz , writing in *Collector's Encyclopedia of American Composition Dolls, Volume II*, the company also sold doll parts to other firms.

15" tall composition shoulder head "Ma-Ma" doll, circa mid 1920s. She is marked on the back of her shoulder head "N.D.Co.," which may stand for Natural Doll Co., a New York firm of the period. The doll has a mohair wig, sleep eyes, and a closed mouth. Her lower arms are composition and her upper arms, body, and legs are cloth. The body includes a crier. She wears her original dress, bonnet, and one piece bloomer underwear. Her shoes and socks are replacements. ($50-75).

24" tall composition shoulder head "Ma-Ma" doll, circa late 1920s. The doll appears to be marked on the back of the shoulder head "ND" inside a large C. It may be that she was also from the Natural Doll Co. She has a mohair wig, sleep eyes, open mouth with teeth, full composition arms, and composition legs to above the knees. Her upper swing legs are cloth as is her body, which includes a crier. She is all original including her peach dress, hat, one piece bloomer underwear, and shoes and socks. ($150).

Dora Petzold

Dora Petzold began making doll heads of silk material in Germany, circa 1919. During the 1920s, she produced dolls with heads of molded composition and bodies and limbs of cloth. The facial features were hand painted with oil paints. The trademark (a "DP" with a girl in a circle) was registered in Germany in 1924, according to the Colemans writing in *The Collector's Encyclopedia of Dolls*.

25" tall composition head doll by Dora Petzold of Berlin, Germany, circa 1920s. The doll has a mohair wig, painted features with a closed mouth, and a cloth body, arms, and legs. The outfit appears to be original. The trademark stamped on the cloth bodies is "DORA PETZOLD/Registered/Trademark/Doll/Germany." ($1,200-1,300). *Doll and photograph from the collection of Jean M. Grout.*

Jessie McCutcheon Raleigh

Jessie McCutcheon Raleigh was in the doll business in the Chicago area for only a short time, from 1916-1920. Raleigh had special composition developed for her dolls and the first line of dolls was issued circa 1917. The faces were hand painted. By 1919, sleep eyes and wigs had been added to the all composition doll. Other dolls with wigs and sleep eyes came with composition shoulder heads, lower arms, and legs, plus cloth bodies and upper arms and legs. The dolls did not seem to be successful, perhaps because they were so heavy, and the company was sold circa 1920.

Butler Bros. advertised the Raleigh dolls in 1918 and 1919. The Lamp Post Antiques, in Colton, California, reproduced old catalog ads in a publication called *Dolls Toys - Miniatures 1900-1930*, circa late 1960s or early 1970s. Pictured in the publication are five pages of Raleigh dolls from the original Butler Bros. catalogs. Also reproduced is a page from a Sears catalog circa 1917-1918 that pictures two dolls that appear to be from the first line of Raleigh dolls. Both all composition dolls are nude in the picture. The baby is 18" tall and the girl is 19" in height (usually listed as 18.5"). The description states that the dolls were modeled from real children by a famous artist. The description continues "Molded entirely of light weight unbreakable composition… Each doll fitted with a very fine quality chemise." Each of the dolls weighed five pounds and cost $3.95.

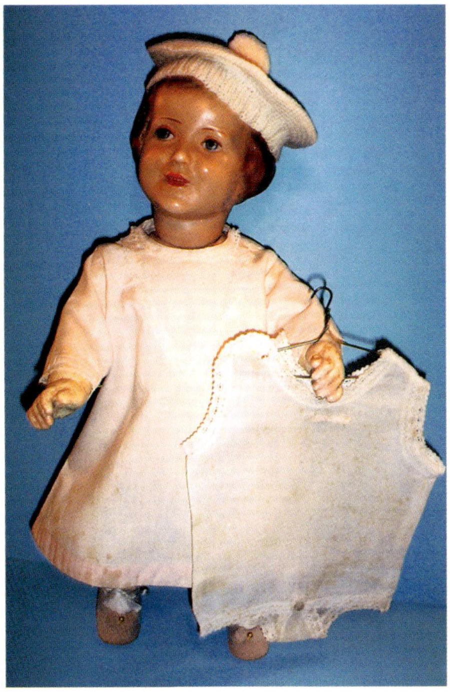

18.5" tall Jessie McCutcheon Raleigh all composition doll, circa 1918. She has molded hair, painted features, and is spring jointed at the shoulder and hips. She is wearing her original dress and holds her original underwear on a hanger. Her hat, shoes, and socks have been added. A similar Raleigh doll was advertised in the Butler Bros. wholesale catalog for that year. The dolls came in sizes of 11.5", 13.5", and 18.5". Many of the dolls were barefoot and dressed only in underwear. Prices ranged from $1.80 to $3.25 each. Baby dolls and dolls with wigs were also available. (With wear to composition, $300+).

A close-up of the Raleigh head shows the unusual painted face of the 1918 dolls. They were modeled from real children but little girls probably preferred dolls with wigs and sleep eyes to provide more play value. The dolls of 1919 included these new features.

Raleigh composition dolls advertisement in *Toys and Novelties* magazine in September 1919. Pictured are three of the newly designed Raleigh Sextette dolls. They were 22" tall, had sleep eyes and real hair, and included composition shoulder heads, lower arms and lower legs, with cloth bodies, upper arms, and upper legs. These dolls were much more elaborately dressed than most of the Raleigh line. *Advertisement from the collection of Marge Meisinger.*

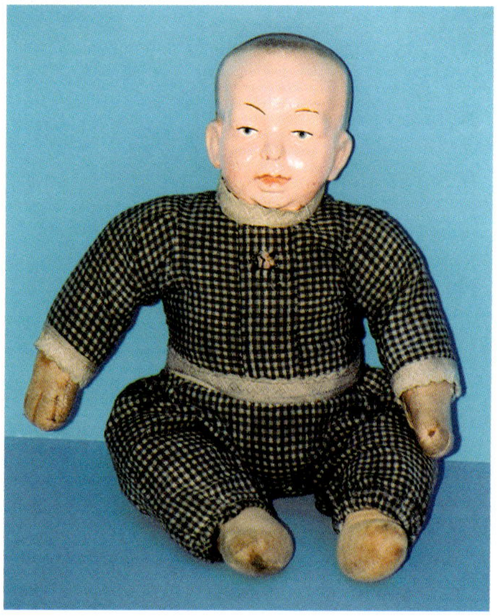

12" tall composition flange head doll, circa 1911-1912. The doll is marked on the back of the neck "WA" followed by a backward P. The maker is unknown. He looks very similar to a "Romper Baby" advertised by Hahn & Amberg in 1910 in *Playthings* magazine. Their doll included cloth hands with the large stitches to indicate fingers as this doll has. It is known that Amberg imported some of their dolls from Germany. The pictured doll is one of several composition models from the period that were based on the Kämmer & Reinhardt #100 bisque baby (the so-called Kaiser baby). This one has painted hair and eyes, closed mouth, and a cloth body. He is jointed at the shoulders and hips with inside joints. He is wearing his original romper suit, which looks like the rompers worn by Horsman's Baby Bumps Dolls. ($150).

11" tall composition flange head doll, circa 1911-12. Although she looks very much like a Campbell Kid, her hair is parted on the opposite side. She has molded painted hair, large painted eyes to the side, a closed mouth, composition lower arms and hands, and cloth body, upper arms, and legs. Her body is quite "fat" and her legs are short. The legs and feet are made of sateen cloth with brown used to make the feet to simulate boots. She is jointed at the shoulders and hips with inside joints. Her dress appears to be original. ($150+).

14" tall unmarked composition flange head doll, circa 1912-1915. He has a finely modeled hairstyle, painted eyes, closed mouth, small composition hands, cloth body, upper arms and legs. The feet are made of black sateen to resemble boots. He is jointed at the hips and shoulders with outside discs. He looks similar to some of the early Effanbee boy dolls and may have been made by that firm. He has been redressed. His "boots" have wear. ($150+).

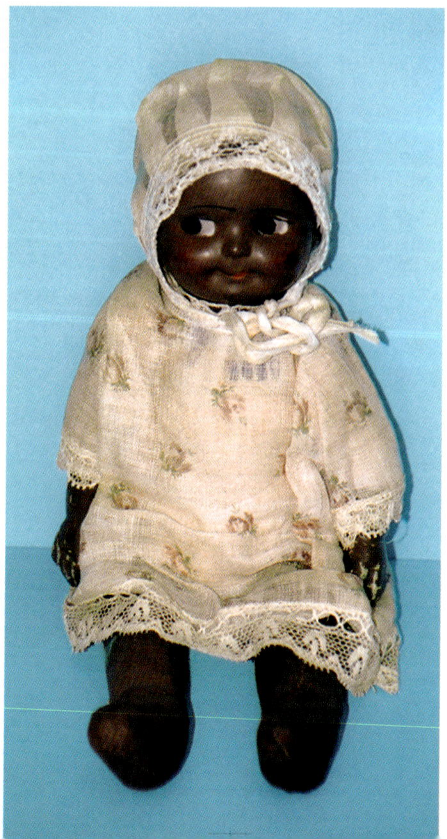

13" composition unmarked flange head doll, circa 1912-1915. She appears to be similar to the doll used on the Ideal "Zu-Zu Kid" in 1916. She has composition gauntlet hands, cloth body, upper arms, and legs, with cloth black feet and striped legs. She is jointed at the shoulders and hips with outside discs on one side and nail like metal on the other. She has been redressed. ($50+).

10" tall unmarked black doll, circa 1918-20. The doll has molded hair and painted features, a composition head and lower hands, and a cloth body, upper arms, and full legs. She is wearing vintage clothing. ($100+). *Doll from the collection of Ellyn McCorkell.*

16" tall unmarked composition flange head doll, circa 1920. She has painted features, a mohair wig, gauntlet composition hands and lower arms, and cloth legs, body, and upper arms. She is wearing her original dress, underwear, hat, shoes, and socks. Her arms and legs are jointed with inside discs. ($100-125).

Pair of unmarked 11" tall composition shoulder head twin babies in their own pillowcase, circa mid 1920s. The dolls have molded and painted hair and features, composition arms, and cloth bodies and legs. ($70 a pair).

This advertisement from the mid 1920s features a set of similar dolls in their pillowcase. The package sold for $1.19. A more expensive pair in a satin pillowcase sold for $3.49.

22" tall unmarked composition shoulder head doll, circa early to mid 1920s. She has molded hair, painted features, composition arms, lower composition legs, and a cloth body and upper legs. She is wearing her original dress, bonnet, underwear, and shoes and socks. ($75+).

12" tall composition shoulder head doll identified only as "Daisy." She may have been used as a premium doll by a magazine company and was given free to subscribers who sold a certain number of new subscriptions to that particular magazine. The doll has molded hair, painted features, composition lower arms and legs, and a cloth body and upper legs. She is wearing her original clothing, including a dress, matching hat (with replaced ribbon), bloomers, shoes and socks. She also has her original tag, which reads "Daisy/Reg. U.S. Pat. Off." A daisy flower is pictured in the corner of the tag. (With paint flaking $50+).

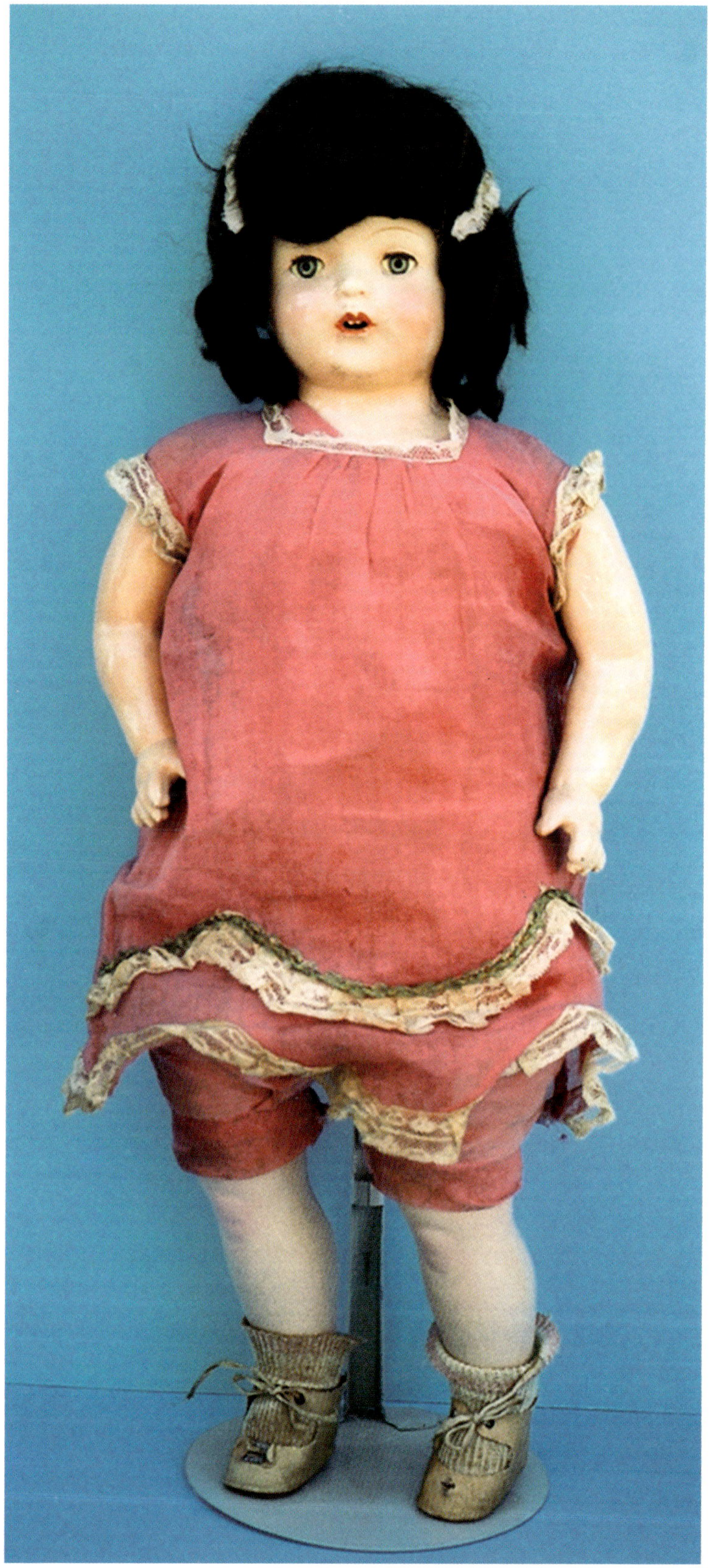

22" tall unmarked composition shoulder head "Ma-Ma" doll, circa late 1920s. She has a mohair wig, sleep eyes, open mouth with two teeth, full composition arms, and composition legs to above the knees. The doll's upper legs and body are cloth, with a crier mounted in her back. She is all original including her dress, one piece underwear, shoes and socks, and hair decoration. ($125+).

Chapter Five

Metal Dolls and Heads

Metal dolls seem to have been overlooked by doll collectors in the past. With the coming of the Internet, that is no longer the case. The German metal shoulder head doll with a scratched head and worn kid or cloth body no longer represents all that is available to today's collectors in the metal doll line. Of course, dolls of this type are still for sale, but in addition, there are dolls with mint metal heads with painted or glass eyes, open or closed mouths, and molded or real hair. Dolls can be purchased made entirely of metal, or with heads and arms or legs of metal. The variety of dolls available on the Internet has turned up examples of metal doll products that even long-time collectors have never seen.

With the wonderful new dolls now for sale, prices have increased along with the newfound interest in fine metal related dolls. The German dolls labeled Juno, Minerva, and Diana are still in demand when in excellent condition, but the more unusual American dolls from the Atlas Doll & Toy Co., the Giebeler-Falk Doll Corp., and the Metal Doll Co. are even more desirable when found in original condition.

Many metal dolls were produced during the later part of the 1800s into the beginning of the twentieth century. The most well-known are the shoulder head dolls from Germany by the firms of Alfred Heller (using the Diana trademark), Karl Standfuss (Juno), and Buschow & Beck (Minerva). The Diana heads were made from approximately 1902 -1908, while the Juno heads were produced beginning in 1904 until perhaps the 1920s. The largest number of these types of heads were manufactured by Buschow & Beck under the Minerva trademark, beginning in 1894 and continuing into the 1920s. The bodies for these heads were usually made of leather with lower arms of bisque, celluloid, or composition. General sizes ranged from approximately 10" to 22" in height.

In 1903, more interesting metal dolls produced by the Metal Doll Co. in Pleasantville, New Jersey, were being marketed in the United States. Dorothy, Elizabeth, and Evelyn Coleman picture an example of one of these dolls in their book, *The Collector's Encyclopedia of Dolls*. The pictured all steel doll is fully spring jointed (including elbow, hand, knee, and foot) and has glass eyes. Three snaps were used to attach a wig to the head. Different wigs could be used to change the appearance of the doll. It stood 17.5" tall.

By 1911, the Sears catalog was featuring metal Minerva shoulder heads with glass eyes and wigs. The heads were priced from 37 cents to 98 cents each, depending on size. Kid bodies could be purchased separately from 45 cents to $1.98 each according to size. They included lower bisque arms and riveted hip joints. Molded hair metal Minerva shoulder heads were also available, priced from 12 cents to 65 cents each.

When fighting broke out in Europe at the beginning of World War I in 1914, toys from Germany were no longer easy to obtain, so companies in the United States began to manufacture different types of metal related dolls to take advantage of the situation. The Atlas Doll & Toy Co. from New York City and the Giebeler-Falk Doll Corp. also located in New York City were two of these firms. After the war concluded in 1918, most of the new firms soon went out of business. By the early 1920s, metal head dolls from Germany were again able to capture a large share of the market. Designs of metal head dolls became more modern in the 1920s and many were featured in the main retail catalogs through the 1920s and to a lesser scale into the 1930s. The makers of most of these dolls are unknown.

One firm whose dolls are easy to identify are those made by A. Buckerer, located in Switzerland during this period. The company produced an unusual line of dolls in 1921 that featured heads, hands, and feet of composition while the body and limbs were made of metal. The dolls were completely ball-jointed. They are marked on their bodies, "Made in Switzerland/Patents Applied For." The dolls were small, mostly in the 8"-12" sizes. Several comic characters were in the line, including Mutt & Jeff and Maggie & Jiggs. Dolls in provincial dress as well as a fireman, aviator, policeman, and others were also available.

Several companies from both Germany and the United States are of special importance as they relate to the manufacture of metal heads or dolls during the twentieth century.

Germany

Karl Standfuss

The Karl Standfuss firm was located in Deuben, Germany beginning in 1904. The company produced both metal and celluloid doll heads as well as celluloid dolls and parts. They used the trademark Juno and the metal heads were sometimes marked with a crown. The firm continued business into the 1920s, manufacturing collectible celluloid dolls including Kewpies and Bye-Lo Babies.

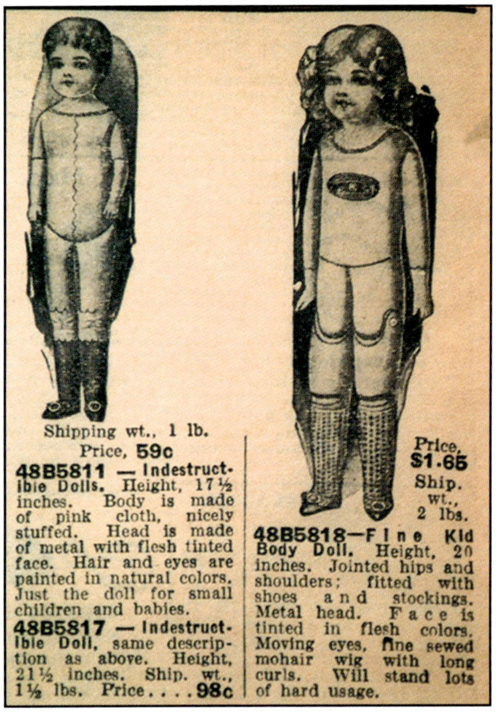

Advertisement in the Montgomery Ward 1916 catalog, which pictured two types of metal head dolls. The cheapest one with painted hair and features sold for 59 cents and included a body made of pink cloth. The doll measured 17.5" tall. Also available was a metal head doll with a kid body, sleep eyes, and a mohair wig. She was priced at $1.65 and was 20" tall.

21" tall metal shoulder head doll that looks like a bisque example until she is more closely examined. She has large blown glass eyes, open mouth with teeth, and a replaced mohair wig. Her body is kid with bisque lower arms. She is marked "6" on her back and a large crown is embossed on her breast plate. Although the crown is known to be a mark used by Karl Standfuss, this doll head seems to be superior to others made by the firm. She has been redressed. ($225+).

10" tall doll with metal shoulder head marked "Juno," made by the German firm of Karl Standfuss. The hair is molded and the features are painted. She has a commercially made cloth body with cloth feet and white bisque hands. A crown is pictured on the front of the head along with the "Juno" mark. Germany is embossed on the back of the head. The head is similar to the heads made by the German firm of Buschow & Beck, whose trademark was "Minerva." ($50 in this condition).

Buschow & Beck

The Buschow & Beck company was located in Germany and began using Minerva as a trademark in 1894. According to the Colemans writing in their book, *The Collector's Encyclopedia of Dolls*, Minerva was registered as a trademark in the United States in 1901. Most dolls were also marked with a helmet design. The firm produced celluloid doll heads and dolls in addition to their more famous metal heads, which remained popular into the 1920s.

20" and 18" tall metal shoulder head dolls made by Buschow & Beck. Both are marked "Minerva" on the front of their shoulder plates and carry the "Helmet" design used by the firm as a trademark. The larger doll is marked "Germany/6 1/0" on her back and the smaller doll is marked "6/Germany." The firm produced many similar hairstyles on its metal heads but there are differences in their wide variety of dolls. The larger doll has an open mouth with teeth while the smaller doll's mouth is closed. Both have painted eyes. Most of the earlier dolls carry the red line above the eye as did the china doll heads produced by other firms. The larger doll has a kid body with jointed kid legs and composition lower arms. The smaller model has bisque lower arms and cloth lower legs and feet. Both dolls have been redressed. (With excellent heads $175 each).

16" tall doll with metal shoulder head marked "Minerva," made by the German firm of Buschow & Beck. She has molded painted hair, painted eyes, and a closed mouth. The head has been mounted on a kid body with bisque hands and lower legs and feet of cloth. Since the heads were also sold alone, it is difficult to know if the body was original to the doll. She is marked "Minerva" on the front of the shoulder plate and "Germany" on the back. The company sold metal dolls with the Minerva trademark from 1894 into the 1920s according to the Colemans writing in their book, *The Collector's Encyclopedia of Dolls*. She has been redressed. ($125+).

12" tall metal shoulder head doll marked "Minerva" on her front and "Germany" on her back. She has molded hair, painted eyes, and a closed mouth. Her body is imitation kid with lower arms of composition and lower legs and feet of cloth. She has been redressed. ($65-75).

12" tall metal shoulder head doll with molded hair similar to the previous example. The boy, however, has glass eyes. He is also marked "Minerva" on the front of his shoulder plate and "Germany" on his back. He has a closed mouth, and the head has been mounted on a commercial cloth body. His head is in excellent condition. The doll has been redressed. ($125).

21" tall Minerva metal shoulder head doll with molded hair, blown glass eyes, open mouth, and teeth. She has a leather body with lower bisque arms, and cloth lower legs and feet. She is marked "Minerva" along with the Helmet trademark on the front of her shoulder plate and "Germany" on the back. Unlike the previous boy doll, this is the condition found most often in metal head dolls. ($40-50).

25" tall metal shoulder head doll marked "Minerva" on her front shoulder plate and "Germany" on her back. This head is also a quality head with the bisque look. She has large brown glass eyes, open mouth with teeth, and a replaced wig. The doll has a leather body with bisque lower arms. She has been redressed. ($200+)

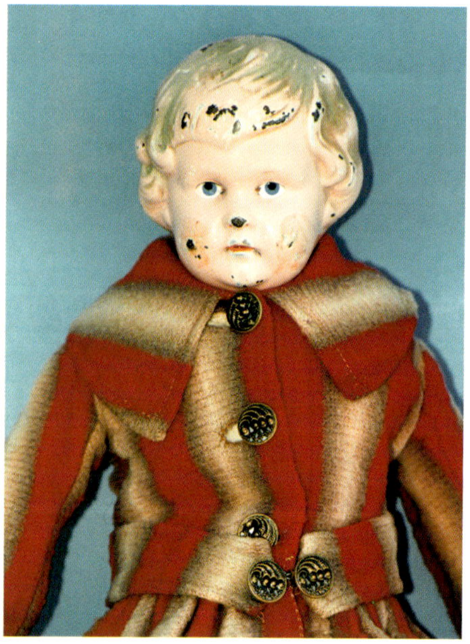

19" tall Minerva metal shoulder head doll. Although the doll is not in very good condition, she is included because she has a very unusual hairstyle for a Minerva doll. She appears to have been designed to represent a younger child. She is marked "Minerva" as well as "Germany 170/6 13 cm." The "13cm" mark is also unusual and perhaps this head dates from the 1920s. She has a worn leather body and is dressed in a vintage coat. A hat matching the coat also came with this doll. ($65-75).

Another example of a 16" marked Minerva metal shoulder head doll has a different look to her face. She wears a mohair wig, has glass eyes, an open mouth, and teeth. Her leather body is jointed at the knees and hips. Her lower arms are celluloid. She has been redressed. ($150-175).

Advertisement from the Fall and Winter Sears catalog of 1923. Four different dolls with metal heads are pictured. The cheapest dolls sold for 59 cents and were 11.75" tall. They were called "Our 'Knock Out' Grade" and were made with Minerva metal heads, painted eyes, and molded hair. The "Wearwell" brand dolls included Minerva metal heads, glass eyes, mohair wigs, and pink cloth bodies with wood lower arms. They ranged in size from 15" to 22.5" tall and were priced from 99 cents to $1.98 each. The most expensive metal head dolls featured Minerva heads with sleep eyes, wigs, and imitation kid bodies with celluloid arms and hands. They came in sizes from 16" to 23.5" tall and ranged in price from $2.19 to $3.98 each.

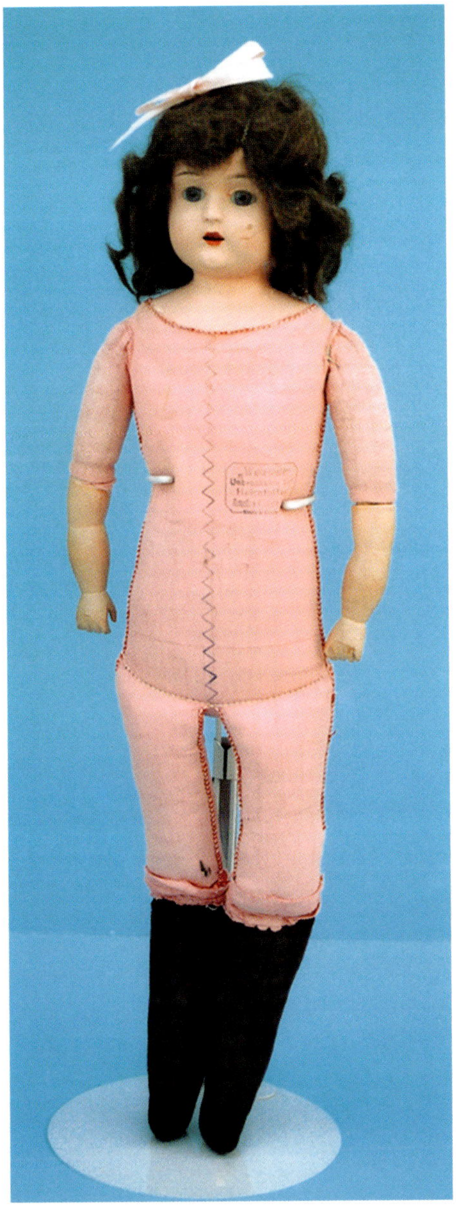

20" tall "Wearwell" brand metal shoulder head doll similar to the ones pictured in the 1923 Sears ad. Her body is stamped "Wearwell/Unbreakable Doll/Hairstuffed/Minerva Head/Made in Germany." She has lower arms made of wood and a crier in her back. ($175+).

Close-up of the "Wearwell" brand metal shoulder head doll featured in the 1923 Sears ad. The advertisement identifies her as having a Minerva head. She has her original mohair wig, glass eyes, open mouth and teeth. She has been redressed.

16" tall metal shoulder head doll marked "Minerva" on the front of her shoulder plate and "Germany" on the back. The wigged Minerva dolls are usually harder to find than those with molded hair. This example has her original mohair wig, glass eyes, open mouth, and teeth. She was made with a leather body jointed at the elbows, knees, hips, and shoulders. Her lower arms are bisque and her lower legs are composition. She has been redressed. ($175-200).

United States

Giebeler-Falk Doll Corp.

The Giebeler-Falk Doll Corp. was a New York firm that began producing an unusual metal head doll in 1919. The firm advertised the doll frequently during the year. Although the doll always included an aluminum socket head, the ads changed during the year as more metal was added to the doll's construction. Aluminum hands were included in advertising in April, and by September aluminum feet had also been incorporated into the doll's construction. The doll had a wood ball-jointed body and metal sleep eyes.

According to the Colemans' information in their book, a phonograph doll with an aluminum body and head was also produced by the firm during this same period.

25" tall Giebeler-Falk doll with an aluminum socket head and hands on a wood ball-jointed body. The feet on this doll appear to be wood (as advertised in April 1919 ad). The doll has metal sleep eyes, an open mouth with teeth, and her original mohair wig. She is wearing a tagged chemise and what appears to be her original shoes and socks. She is marked on the back of her head with a star that includes the letter "G" inside, the number "25," and the letters "US PAT." Her chemise tag reads "Gie-Fa Aluminum Head and Hands, Guaranteed Unbreakable, New York, N.Y." ($225+).

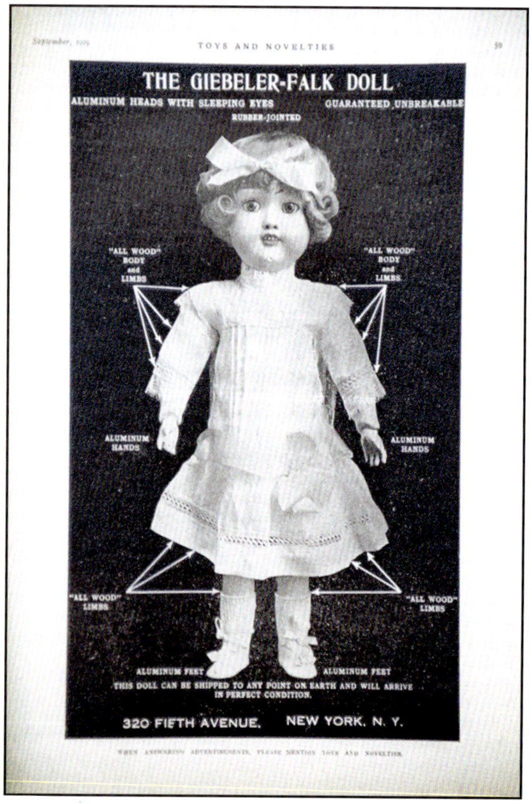

Advertisement from the Giebeler-Falk Doll Corp. that appeared in *Toys and Novelties* magazine of September 1919. This new Giebeler-Falk doll was made with an aluminum socket head and aluminum hands and feet. Her ball-jointed limbs were wood. The New York based firm apparently was only in business from 1918-1921. *Magazine from the collection of Marge Meisinger.*

Atlas Doll & Toy Co.

The Atlas Doll & Toy Co. of New York City was another firm that was producing metal dolls and parts during this period. The firm manufactured dolls and parts of stamped metal. The "Hug Me Kid" doll advertised in 1919 was called "A New creation, the All-Metal Doll." Although the doll was produced to rival the similar composition novelty dolls, today's collectors usually think of the doll as a baby doll. Other dolls marketed by the firm included dolls with metal heads on cloth bodies with composition arms and metal legs, as well as dolls with metal heads on cloth bodies with metal or composition arms and cloth legs. Other combinations might also have been produced. The firm moved to Baltimore, Maryland in 1920 and continued producing dolls in that location for several more years.

Advertisement from the Atlas Doll & Toy Co. of New York City that appeared in *Toys and Novelties* magazine of September 1919. The ad featured a new all-metal doll called "Hug Me Kid." The pictured doll included eyes facing the side and a wig. The copy also stated that the firm offered dolls with metal heads and stuffed bodies. *Magazine from the collection of Marge Meisinger.*

12" tall, all-metal doll like the Atlas doll pictured in the *Toys and Novelties* ad. It has been redressed and includes a replaced wig. ($125+).

The all-metal Atlas doll has sleep eyes looking to the side, closed mouth, and molded hair under the wig. The head has a small hole in the top for wig placement. The doll is jointed at the neck, shoulders, and hips.

A similar metal head little girl doll also appears to be an Atlas Doll & Toy product. She is wearing her original clothing that includes a dress, bonnet, underpants, slip, and socks. She is in very good condition with her original paint still intact. (200+). *Private Collection.*

Instead of being an all-metal doll, this little girl has a metal shoulder head, metal arms, and a cloth body and legs. She has molded hair, painted eyes, a closed mouth, and is unmarked. Her arms are like the metal arms in the previous baby doll. *Private Collection.*

Atlas baby doll with a metal flange head, composition full arms, cloth body, and metal legs. These legs are modeled like the legs on the earlier pictured all-metal baby doll, except that they are attached to a cloth body. There is no marking on the doll. A no longer working crier is in her cloth body. She has molded hair, sleep eyes, and a closed mouth. The design of her head appears to be modeled on the Bye-Lo doll popular in the mid 1920s. She is dressed in handmade old clothing ($200+).

Advertisement from the Fall and Winter M.W. Savage catalog from 1929-1930. The Minneapolis firm featured three metal head dolls in their catalog that year. One is similar to the Bye-Lo baby with sleep eyes. Prices range from 59 cents to 79 cents each. The maker is not known. *From the collection of Marge Meisinger.*

13" tall additional baby metal head dolls that have the look of the Atlas Doll & Toy Co. but are not marked. The doll on the left features a metal head fitted with a separate metal shoulder piece while the doll on the right has a shoulder head. Both dolls have molded hair, sleep eyes, closed mouths, and are on bodies with composition arms and cloth bodies. The doll on the left has composition lower legs and the doll on the right has cloth legs. Both may be products of the Atlas Doll & Toy Co. ($50-100 each).

J. Chein & Co.

J. Chein & Co. was founded in 1903 by Julius Chein in New York City. The firm moved to Harrison, New Jersey in 1912. Although it is best known as a maker of wind-up toys and other boy toys, the firm did produce at least one line of clearly marked dolls with metal heads. The dolls, probably circa 1930s, are marked under the metal heads on a flat bottom piece. If a mark cannot be found, ears drawn onto the sides of the heads can help to identify the doll. The dolls have composition arms and lower legs with cloth bodies that contain a crier box.

15" tall metal head doll manufactured by J. Chein & Co., circa 1930s. She has a flange neck attached to a cloth body with nearly full composition arms and lower composition swing legs. This elusive doll may date to the mid to late 1930s but is included here because it is such an unusual metal head doll that it is seldom seen. The head is made in two pieces and joined together with metal tabs. The ears are drawn on the head, which gives the doll an incomplete look. The hair is molded and the mouth is closed. The doll has sleeping tin eyes. She has been redressed. ($125).

Pictured is the bottom of an identical Chein metal head that is in poor condition. The company marking is on the underside of the head in a shield and reads "J. Chein & Co./Made in U.S.A." On the front of the neck at the bottom is "Patent Pending." The pictured crier box was mounted in the doll's back.

Unidentified Dolls

Advertisement in the M.W. Savage Fall & Winter catalog of 1929-1930 picturing the new line of more modern dolls with unbreakable metal heads. They include a Baby Joan doll in 13" and 17" sizes. She has a metal head, sleep eyes, molded hair, composition arms, cloth body with crier, and cloth legs. She sold for $1.19 for the smaller version and $1.98 for the larger doll. A "Ma-Ma" doll was also available in 20.5" and 22" sizes. She featured a metal head, sleep eyes, mohair wig, cloth body, and composition arms and legs "in the new slender style." This model was priced at $2.25 for the smaller version and $2.75 for the larger model. The third doll came in a 15" or a 19.5" size. She was made with a metal head, painted hair, sleep eyes, composition arms, cloth body, and legs. *Ad from the collection of Marge Meisinger.*

17" tall, all original little girl doll with a metal head that looks similar to the doll advertised in the M.W. Savage Co. catalog for 1929-30. She has a metal unmarked flange head, sleep eyes, open mouth with teeth, and her original wig. The body is cloth with full composition arms and composition legs to above the knees. She wears her original dress, bonnet, bloomers, slip, shoes, and socks. One thumb is broken. ($150+).

15" baby doll similar to Baby Joan featured in the M.W. Savage ad. She has a metal flange head, molded hair, sleep eyes, composition full arms, composition lower legs, and cloth body (with crier) and upper legs. Her mouth is open and two teeth can be seen inside. She has been redressed and is unmarked. ($65).

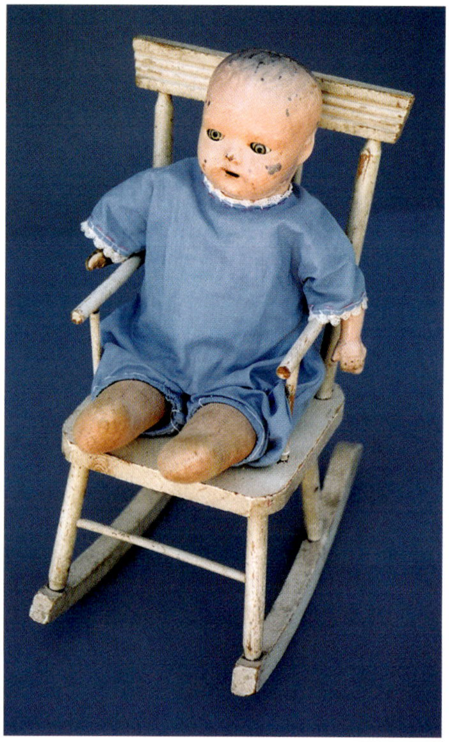

14" tall doll with a very deep metal shoulder head that is similar to the third doll pictured in the Savage catalog. He has lightly molded hair, sleep eyes, open mouth and teeth, outside jointed full composition arms, and cloth legs. His cloth body contains a non-working crier. The doll is dressed in a vintage homemade cotton romper. ($60-75).

20" tall "Ma-Ma" doll, circa 1930, which includes an unmarked metal head, full composition arms, lower composition legs, and a cloth body with a nonworking crier. She has sleep eyes, an open mouth with teeth, and her original mohair wig. The doll has been nicely redressed. ($150+)

13.5" little girl doll, circa 1920s, with an unmarked metal flange head. She has deeply molded hair with a small slot to be used to thread a hair ribbon. The doll has sleep eyes, closed mouth, full composition outside jointed arms, cloth body, full cloth legs, and made on cloth shoes. She has been redressed in a handmade dress from the period that has been decorated with embroidery. ($125+).

As times changed, so did the metal head dolls. A variety of baby and little girl dolls were produced in the 1920s and early 1930s with composition arms and legs and cloth bodies. Some of these dolls included criers to follow the current "Ma-Ma" doll style. The Sears 1927 catalog also pictured a 12" tall baby doll with head, arms, and legs made of metal. The body was cloth. It sold for 99 cents. Most of these dolls were not marked and even when found in original clothing, there is no way to determine the maker.

Although most metal head dolls could not compete with their more attractive bisque counterparts in the collecting field, these dolls did fill an important place in the doll market for decades. They allowed little girls who might never have owned a doll to possess an inexpensive playmate with a metal head mounted on a handmade cloth body or a more expensive leather one made commercially. Perhaps nostalgia for those simpler times is what is driving current collectors to express more interest in the often passed over metal dolls of earlier years.

Chapter Six

Wood Dolls

Schoenhut Dolls

A. Schoenhut & Co. was founded in Philadelphia, Pennsylvania in 1872 by German immigrant Albert Schoenhut. He had come to the United States at the age of seventeen and worked with toy pianos for John Deiser & Sons. Schoenhut was able to make improvements to the pianos then being produced and founded a new company to manufacture his own more sturdy toy pianos. The pianos proved successful and the business grew. Other toys were added to the company's inventory, but it was not until the introduction of the Humpty Dumpty Circus in 1903 that another toy created as much interest and sales as the Schoenhut pianos. Another success for the firm occurred in 1908 when the Rolly Dolly was brought to the market. In 1909, another play set was introduced, called "Teddy's Adventures in Africa." It was based on Theodore Roosevelt's hunting trip to Africa. This toy was sold from 1910-1912 by the Schoenhut firm.

The "All Wood Perfection Art Dolls" did not appear on the market until 1911. An advertisement in *Playthings* magazine in March 1911 pictured several of the new dolls. Included were both boy and girl dolls with carved hair or wigs. According to the Colemans, writing in *The Collector's Encyclopedia of Dolls*, most of the first dolls were designed by Adolph Graziano, an Italian sculptor. The only two dolls in the Schoenhut lines that had specific names were introduced in the first line of dolls in 1911. They were Tootsie-Wootsie, an infant doll with carved hair, and Schnickel-Fritz, another infant-toddler doll also with carved hair plus an open mouth and teeth. Both are rare and very desirable for collectors of Schoenhut dolls. The dolls were made of wood and were jointed with steel springs and swivel joints. Schoenhut advertisements stated that "they can assume and hold any natural pose—running, jumping, dancing, skipping, sitting, or standing." The dolls were painted with oil colors and company advertising advised that the dolls could be washed without injury. The early dolls carried the maker's name impressed on their bodies. It read "Schoenhut Doll/Pat. Jan. 17, '11, U.S.A./ & Foreign Countries."

Although this first line of dolls has much appeal to collectors, the dolls were, after all, made for children to love. Some of them were not attractive to little girls used to the more beautiful bisque head German models. As early as 1912, according to Carol Corson writing in her book, *Schoenhut Dolls*, many of the doll's faces became younger looking and more attractive. All of the dolls continued to have the early leg with no stocking groove.

The Sears catalog for 1912 pictured four of the Schoenhut dolls for sale. Included were a 14" boy with carved hair dressed in a union suit and priced at $1.69, or completely dressed for $2.65. A 14" dressed girl with a mohair wig was $2.98. The popular 14" carved hair girl with a hair band, dressed in a union suit, cost only $1.69. Each one came with the Schoenhut stand, which allowed the dolls to assume a variety of positions by attaching the stand to the holes in the doll's shoes and feet.

Ladies Home Journal for November 1912 carried a small Schoenhut ad that also featured four of the new line of dolls. The dolls pictured included a boy and girl with mohair wigs, a boy with carved hair, and the well-known girl with a bonnet. Many of the early dolls featured intaglio eyes.

The new dolls in the 1912 line also came in a variety of sizes instead of just 16" for the regular dolls and 15" for Tootsie Wootsie and Schnickel-Fritz. Sizes for the new dolls included 14", 16", 19", and 21".

Albert Schoenhut, founder of the A. Schoenhut Co., died in 1912. The Colemans list the Schoenhut sons who then took over the company as Alvert F., Harry E., Gustav A., Theodore C., William G., and Otto F.

In 1913, Harry Schoenhut designed a bent limb baby that included a copyrighted head. It had painted eyes, a closed mouth, and lightly molded hair. The doll body was very different from the bodies of the other dolls. It had spring joints at the neck, shoulders, and hips, but the three piece arms and legs were put together with dowel joints at the wrist, elbow, ankle, and knee. The head was also used on a toddler model with a body like the other Schoenhut dolls. The infant dolls were approximately 13" tall. They were made from 1913 until 1926. Beginning in 1914, the doll could also be purchased with a wig.

Apparently little girls still preferred "dolly faced" German bisque dolls in the mid teens. Since German dolls could no longer be imported because of World War I, the Schoenhut firm decided to meet the challenge by producing a doll called "Miss Dolly" in 1915. This doll had an open-closed mouth with teeth, small ears, painted eyes with stippling in a darker color around their edge, painted upper and lower eyelashes, and a new curlier wig. She came dressed in a union suit with an added ribbon and lace decorated chemise sometimes added. An advertisement from the *Woman's Home Companion* for December 1915 pictures the new Miss Dolly along with the Schoenhut pin with a shield in the middle, which read "Schoenhut/All Wood/Perfection/Art Doll." This doll is most frequently found by collectors. Beginning in 1920, the doll's eyes were made with decals. Miss Dolly continued to be marketed until at least 1926.

A full page Schoenhut ad in *The Ladies Home Journal* of December 1916 pictured the toy pianos, the Humpty-Dumpty Circus, toy cars and trains, and an assortment of dolls. Both carved hair and wigged dolls were featured. The two dolls most prominent in the ad are Miss Dolly (here called "the usual doll-face"), and a character doll labeled "Life-like child face." In the smaller picture, both the doll with the molded hair band

and the girl with the braids around her head with a molded hair bow in the back can be seen. Also pictured are boys with carved hair and many dolls with mohair wigs in both girl and boy models. The ad also states that the Schoenhut toy factory "is one of the largest in the world." Beginning in 1916, the Schoenhut dolls were marked with a decal that read "Schoenhut Doll/Pat. Jan.17/1911/U.S.A."

In 1919, a new "Walking Doll" was added to the Schoenhut line. This was done to meet the competition of other American firms who were also marketing "Walking Dolls." The doll was an infant toddler (using the earlier copyrighted infant head). It was jointed at the neck, shoulders, and hips. The hip joints were a new innovation to allow the doll to "walk." The feet of these dolls did not include the usual holes on the soles because the dolls could not be used with the Schoenhut stands.

The trade magazine *Toys and Novelties* carried a Schoenhut advertisement in September 1919 picturing the new "Schoenhut Walkable Doll." The ad stated that the dolls came in sizes of 11", 14", and 17". The doll needed to be led by the hand to walk. The pictured doll was fitted with a wig. The other doll shown in the ad was "Miss Dolly" dressed in a fancy lace trimmed dress.

Still trying to compete with other doll manufacturers, the Schoenhut firm received patents for movable sleeping eyes in 1921. These eyes were used on many of the "Dolly Faced" dolls and the infant dolls. The infant dolls came in 13" and 15" sizes and were fitted with mohair wigs. A Schoenhut ad from the period pictured a "Miss Dolly" doll with closing eyes in a 17" size. A 14" wigged character doll with closed mouth and painted eyes is also pictured. Another ad from 1921 also mentions closing eyes on the pictured "Miss Dolly." The other dolls include the early carved hair girl with the hair band and an assortment of other wigged dolls. There is no mention of walking dolls.

In 1925, Schoenhut also produced a limited number of unauthorized wood Bye-Lo baby dolls. Photographs of the Bye-Los in Corson's book show baby dolls that look very much like their bisque counterparts. The dolls included cloth bodies and limbs to be used with their heads and hands made of wood.

As the 1920s progressed, new and better composition dolls were being produced by American toymakers. The Schoenhut dolls were expensive to produce with their steel springs and many joints. The Colemans describe a cheaper doll made by Schoenhut beginning in 1924 that was designed to compete with the other toy manufacturers. The dolls were made of Basswood and were jointed with elastic instead of metal springs. In addition, the firm developed new stuffed dolls with cry boxes. These dolls had cloth bodies with wooden heads and hands. They used the infant heads and were 14" or 16" tall. A Schoenhut ad in the November 1926 *Ladies Home Journal* pictured the "Mama" doll along with a wigged infant doll and "Miss Dolly." The eyes were movable or fixed according to the ad.

Carol Corson describes the design of Harry Schoenhut's last doll in her Schoenhut book. The 1930 doll was 13" tall, all composition, with molded hair, and painted features. She was jointed only at the neck, shoulders, and hips. By 1931, Schoenhut no longed listed dolls in their catalogs. Dollhouses and dollhouse furniture that had been introduced in 1917 (houses only) continued to be offered by the firm until its last year of business in 1934.

Given the changes in styles and materials used to make toys in the 1920s (metal for cars and trains and composition for dolls), the A. Schoenhut firm continued to have a hard time competing with the many new toy companies. When the Depression hit in the early 1930s, economic times became even harder. The firm went into bankruptcy in 1935 and the assets were sold. Albert Frederick and his son Frederick Carl received the rights to produce the Schoenhut pianos. They formed the Schoenhut Manufacturing Co. and continued to manufacture other pianos until they went out of business in 1940. Otto Schoenhut, using the name O. Schoenhut Incorporated, continued to produce the pianos until the 1980s. He had earlier been successful with marketing a "Pick-up-Sticks" game.

Many of the early Schoenhut toys have lasted more than one hundred years and it is surprising how many of the playthings are still in good to excellent condition. Most of the dolls have lost paint and/or their original wigs but their spring joints still work beautifully. Although most collectors prefer all original and excellent examples of dolls for their collections, the unusual and carved hair examples are still desirable even with some wear and paint touch-ups. With nearly two decades worth of marketed dolls to be found and the Internet offering a wide range of sellers, current collectors have the opportunity to turn up more Schoenhut treasures than ever. (See the "Comic and Celebrity Dolls" chapter for more Schoenhut products.)

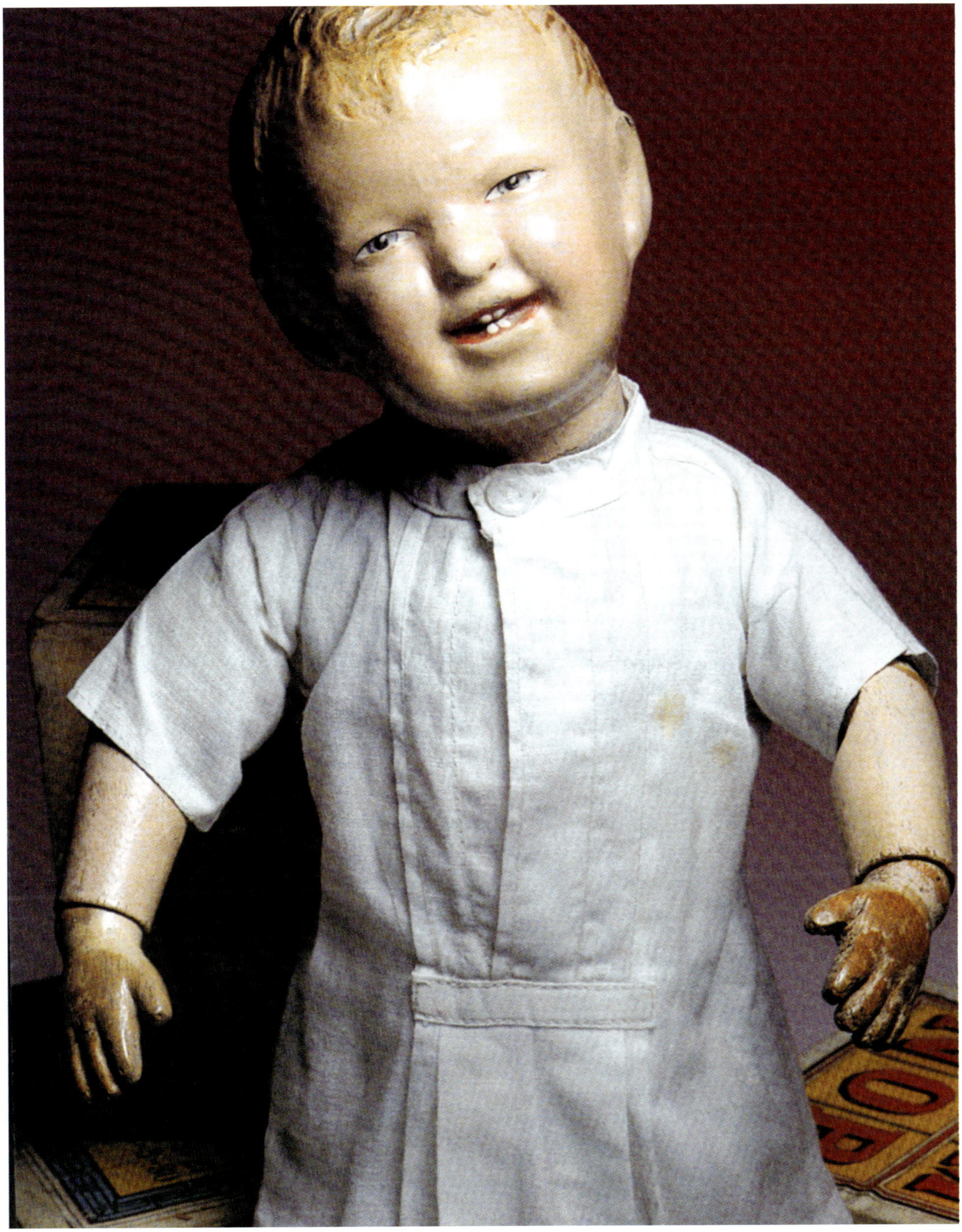

15" tall Schnickel-Fritz all wood Schoenhut toddler type doll first made in 1911 in the first line of dolls. He was designed by Adolph Graziano, an Italian sculptor. He is one of only two dolls in the Schoenhut line of dolls ever given a specific name. (The other one that was also designed by Graziano for the 1911 line of dolls was called Tootsie-Wootsie. It was more of an infant doll.) All of the dolls were jointed with steel springs and swivel joints. This doll, like all of the early dolls, carries the maker's name impressed on his body. It reads "Schoenhut Doll/Pat. Jan. 17, '11, U.S.A./& Foreign Countries." This doll has a carved wooden socket head, deeply-carved and painted wavy blonde hair, painted features, blue eyes, and an open-closed mouth in a wide smile with two upper and lower teeth. He wears his original cotton chemise, newer underwear and shoes and socks. This model doll originally sold for $3.00 undressed and $5.35 wearing an infant dress. He also came dressed in rompers. ($3,500+). *Courtesy of Frasher's Doll Auctions, Inc., Oak Grove, Missouri.*

A. Schoenhut Co. advertisement from the *Ladies Home Journal* of November 1912. The Philadelphia based firm first began producing dolls in 1911 but the company had been in business since 1872. Pictured in this ad is the bonnet head doll (16/106), a carved hair boy, and a boy and girl with wigs. The earliest dolls were 16" tall except for Schnickel-Fritz and Tootsie-Wootsie who were 15" in height. The bonnet doll was carried through 1916 in 14", 16", and 19" sizes.

16" tall early Schoenhut boy doll with carved hair (16/201), circa 1912. He has side-parted, carved and painted blonde hair, painted features, intaglio blue eyes, closed mouth, and spring-jointed body. He carries the incised Schoenhut mark on his back. He has some wear including a broken thumb. ($1,100-1,500). *Courtesy of Frasher's Doll Auctions, Inc., Oak Grove, Missouri.*

16" tall Schoenhut girl doll with carved hair and a headband (#105). This doll was first produced in 1912 but remained in the line of dolls through the early 1920s. She has a carved wooden socket head, molded and painted short brown hair, and a blue headband with a bow in back. The doll has painted features, intaglio blue eyes, closed mouth, and a carved wooden spring-jointed body. She has been redressed. She is marked with an oval sticker on the back. The other pictured doll is 19" tall and is an older Schoenhut model, made circa 1915-1918. This doll carries the incised Schoenhut mark on her back. She has painted features, intaglio blue eyes, closed mouth, original brown human hair wig, and carved wooden spring-jointed body. She is wearing a blue and white cotton Red Cross nurse's uniform. She is in excellent condition and retains her original finish. (Hair band doll $1,500+, Pouty $1,000+). *Courtesy of Frasher's Doll Auctions, Inc., Oak Grove, Missouri.*

21" tall, carved hair girl doll with braids (102) model, circa 1912-1916. She has brown hair accented by a pink bow at the back, painted features, intaglio brown eyes, eyeliner, feathered brows, closed mouth, and a spring-jointed body. She has the incised Schoenhut mark on her back. She is wearing her original dress and union suit. Her shoes and stockings have been replaced. She was made in 14", 16", and 19", and 21" sizes. She was one of the firm's most popular dolls and is also a favorite for today's collectors. Also pictured is a Schoenhut Rolly Dolly, 11" tall. (Doll $1,900-2,200, Rolly Dolly $400+). *Courtesy of Frasher's Doll Auctions, Inc., Oak Grove, Missouri.*

13" tall infant doll on a bent limb baby body designed by Harry Schoenhut in 1913. The doll has molded hair, painted eyes, and a closed mouth. She has spring joints at the neck, shoulders, and hips, and three-piece arms and legs put together with dowel joints at the wrist, elbow, ankle, and knee. The back of the head is marked with a circle decal that reads "H.E. Schoenhut/C/1913." Another decal on her back reads "Schoenhut Doll/Pat. Jan 17th 1911/U.S.A." She is wearing her original chemise that has been made into a dress with an added bonnet. (With wear $250).

Schoenhut ad in the *Woman's Home Companion* for December 1915, advertising the new "Miss Dolly" doll. This doll was made to compete with the prettier bisque head dolls. She has an open-closed mouth with teeth, painted eyes with painted eyelashes above and below her eyes, and a new styled curly wig. This doll was the biggest selling Schoenhut doll and was kept in the line until at least 1926. The Schoenhut pin is also pictured in the ad.

21" tall early Miss Dolly doll. She has blue eyes, open-closed mouth with teeth, painted eyelashes above and below her eyes, and red dots in the corner of her eyes and in her nostrils. The ears on these dolls are small for the size of the dolls. This one has the incised Schoenhut mark on her back. Her body is in excellent condition but her face has had some touch up and her wig has been replaced. ($300+).

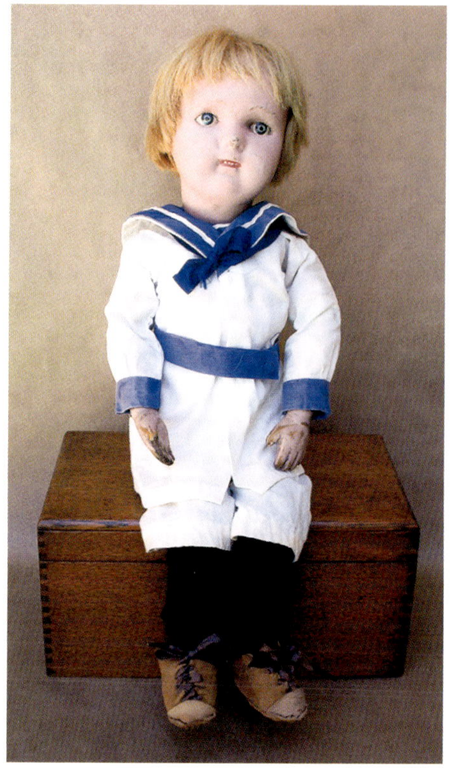

16" tall dolly faced doll with blue decal eyes, circa 1920-1924. The doll has four painted teeth and an open-closed mouth. His wig has been replaced and paint touch ups have been made on his face. The body paint is original. He is marked on the back with the Schoenhut decal mark. The doll is wearing a vintage homemade sailor suit. ($300+). *Photograph and doll from the collection of Jeff Zillner.*

19" tall character face doll thought to be #308, which was also made as a boy as #405, according to Carol Corson writing in her *Schoenhut Dolls* book. This doll is probably the most easily found character doll as it was produced for around twelve years. It also came in a 14" size and in a 22" size beginning around 1917. She has deeply incised intaglio eyes, lower lip with a dip in the center, and a darker pink line between the lips. The lips curl up at the corners if the paint is still visible. Her upper eyelids are deeply molded. This doll is marked with the decal indicating she was made after 1916. Her wig has probably been replaced along with her shoes and socks. She wears a union suit that may be original along with a slip trimmed in ribbon and lace, which may also be a Schoenhut product. The old pink, black, and white gingham dress, trimmed in lace and eyelet, is of the Schoenhut style with a low waist. It is not known if it is original. ($800+).

The A. Schoenhut Co. placed a full page ad in the *Ladies Home Journal* for December 1916. It pictures a doll showing the spring joints that hold her together. Also featured are a "Miss Dolly" doll in her union suit and a character face doll. At the top of the page, many different dolls from the Schoenhut line are shown. Included are those with carved hair in both girl (hair band and braids) and boy styles. A variety of wigged dolls are also pictured. In addition, boys' toys and a Schoenhut piano are featured in the ad.

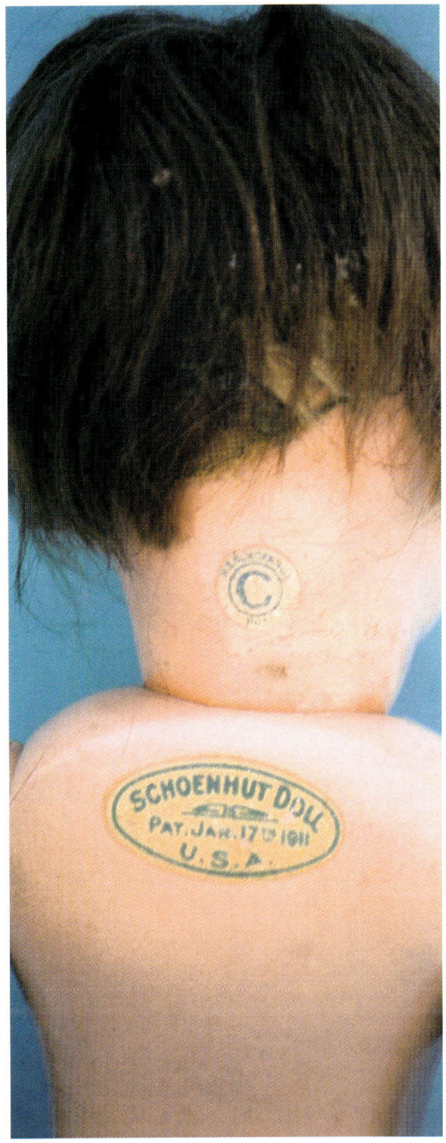

The dolls were marked with a decal mark beginning circa 1916. The oval decal stated "Schoenhut Doll/Pat. Jan. 17th 1911/U.S.A." This walking doll head is also marked "H.E. Schoenhut ©1913."

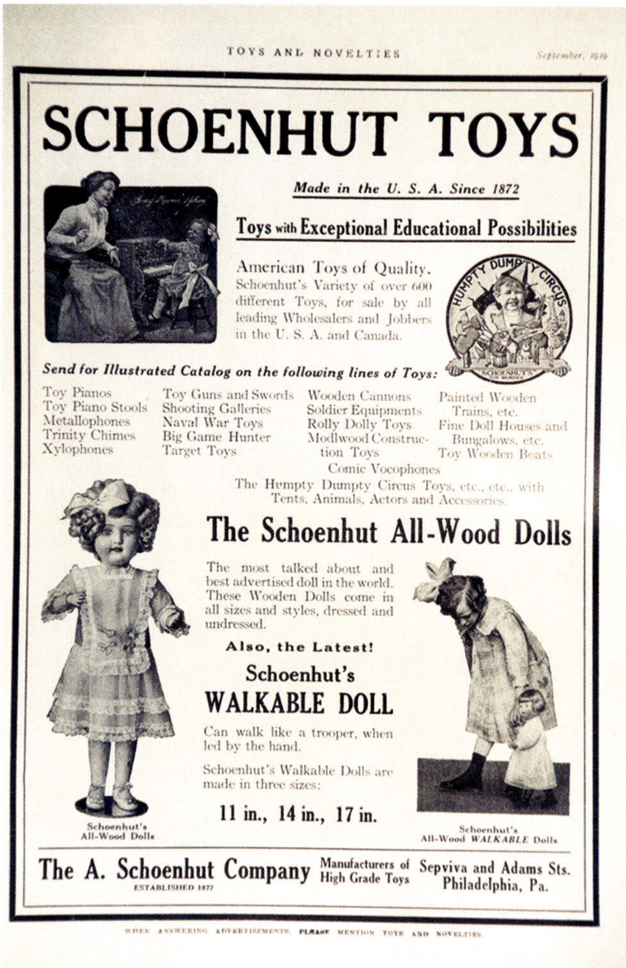

In 1919, a new "Walking Doll" was added to the Schoenhut line. *Toys and Novelties* for September 1919 carried a Schoenhut advertisement picturing the new "Schoenhut Walkable Doll." The ad stated that the dolls came in sizes of 11", 14", and 17". The pictured doll was fitted with a wig. The other doll shown in the ad was "Miss Dolly," dressed in a fancy lace trimmed dress. *Advertisement from the collection of Marge Meisinger.*

Schoenhut ad circa 1921 featuring the dolly faced doll with sleep eyes. She was 17" tall and #317C. She retained the curly wig first introduced on "Miss Dolly" in 1915. The other doll pictured was #107W in a 14" size. She had stationary eyes and a bob cut wig. The Schoenhut pin is pictured on the front of her dress. This is an example of the Toddler doll that used the infant head.

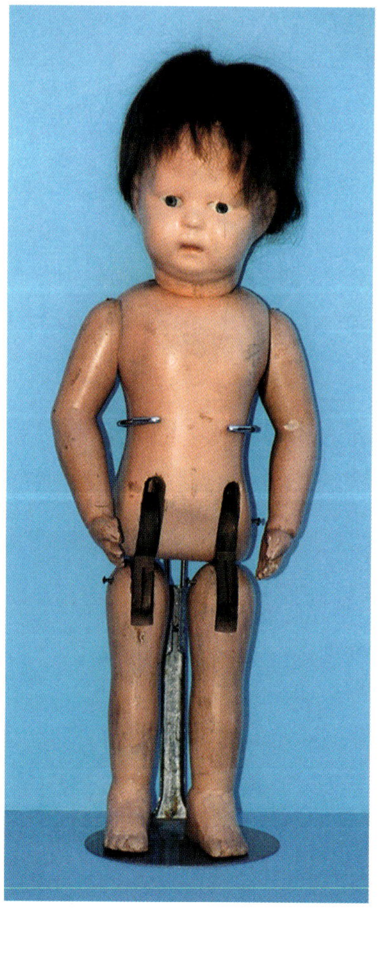

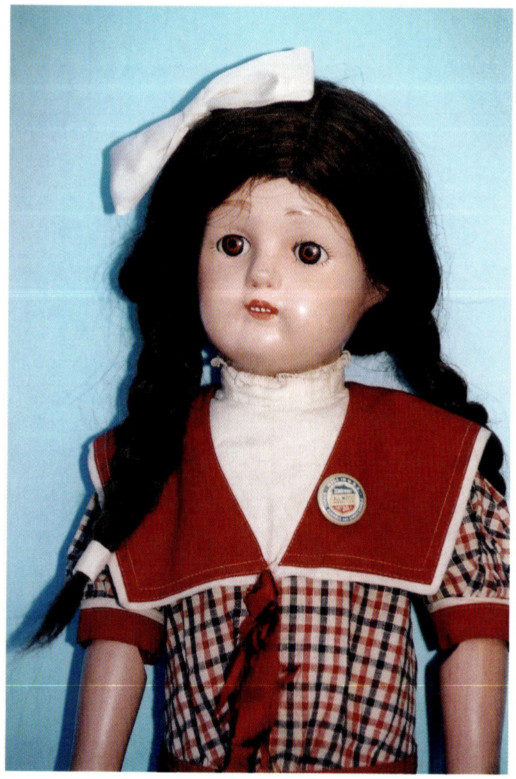

Right:
11" tall walking doll, circa 1919, using the infant head marked with the infant circle decal on the back of the head. The back of the doll bears the regular Schoenhut doll decal. The arms and neck are spring jointed. The legs have a special walking joint. The doll has painted eyes, a closed mouth, and his original wig. ($350-400).

Far Right:
Marked sleeping eye Schoenhut doll first made in 1921 and continued in production until 1928. The dolls were the earlier "Miss Dolly" dolls with added sleep eyes to make the dolls even more like other composition and bisque head dolls then on the market. The doll is 22" tall with brown sleep eyes. Her wig may have been replaced. She is wearing what appears to be an original Schoenhut union suit and a vintage dress made in the Schoenhut manner that may be original. Her paint is in excellent condition. She has an open-closed mouth with four painted teeth. The Schoenhut pin is a replica. She has the Schoenhut decal on her back. (Doll, $600-700).

Schoenhut doll ad from December 1921, which also mentions sleep eyes. A "Miss Dolly" with sleep eyes is pictured. Wigs are described as being made of mohair in blonde or tosca. Only one carved hair doll is pictured (with hair band). The other dolls shown are wigged girls except for one wigged boy.

Illustrations of Dresses for the "Schoenhut Doll"

We manufacture a large line of fashionable, attractive and serviceable Doll Dresses for both Girl and Boy Dolls. Our Doll Dresses are all fac-similes of real children's dresses made from good material and are of first-class workmanship. In ordering Doll Dresses always prefix the size of the Doll to the style number of the Dress; for instance, you want Dress No. 567 for 14-inch Doll, ask for 14/567. With each Girl Doll Dress we include White Lawn drawers and petticoat with bodice trimmed with lace. Slippers and Stockings are furnished with the Dolls and are not furnished with Dresses bought separately.

No. 365 No. 566 No. 567 No. 568 No. 569

No. 530 No. 582 No. 583 No. 584

No. 812 No. 813 No. 818 No. 819

The suits made for the dressed Boy Dolls are made exactly like real boy suits and are strong and durable. Each dressed Boy Doll has on a Union Suit undergarment underneath his suit.

14

Page from a Schoenhut doll catalog, circa 1916, picturing thirteen different outfits that could be purchased for Schoenhut boy and girl dolls. The dresses sold from $1.35 to $3.00 each, depending on size and style. The frilly white dresses were the most expensive. The boys' suits were priced from $1.25 to $3.00 each.

Schoenhut advertisement from the *Ladies Home Journal* of November 1926. The pictured dolls include Miss Dolly, a baby doll with curved arms and legs, and the "Ma-Ma" doll with a wood infant head and hands, plus a cloth body, legs, and arms. All three dolls are wigged. The ad copy read that dolls could be purchased with movable or fixed eyes and with carved hair or mohair wigs. *Ad courtesy of Marilyn Pittman.*

The same Schoenhut catalog also offered a five-piece set of doll furniture that included a round table and four matching chairs. The outfit came in two different sizes. The cheapest set was designed to fit 14"-16" dolls and sold for $1.50. The larger set was suitable for 19" and 21" dolls and cost $2.50. The catalog also included shoes, union suits, stockings, parts, wigs, and repainting and refinishing work. An entire doll could be refinished for $1.15 for a 14" doll. Doll prices listed in the catalog included $2.75 for 14" carved hair dolls and $4.00 for 16" sizes. Wigged dolls ranged from $3.00 to $3.50 for 14" dolls and $3.50 to $3.85 for 16" dolls. All of these models were dressed in union suits. Dressed dolls could cost up to $8.00 each. They ranged in size from 14" to 22" tall. Infant dolls were priced from $3.00 to $6.85, depending on size, wig, and clothing. The largest infant, with "nature arms and legs," was 11" tall. The smallest measured 9" high.

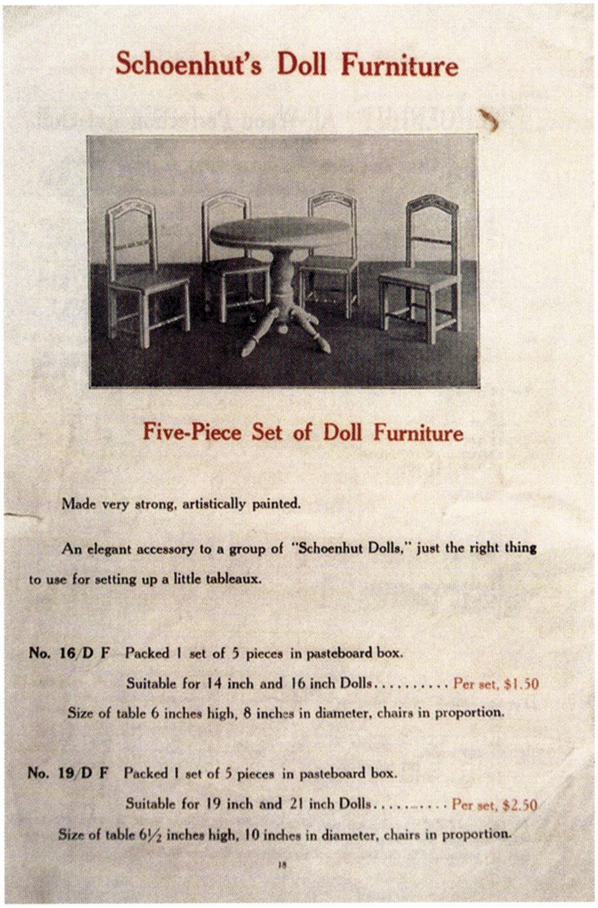

Schoenhut's Doll Furniture

Five-Piece Set of Doll Furniture

Made very strong, artistically painted.

An elegant accessory to a group of "Schoenhut Dolls," just the right thing to use for setting up a little tableaux.

No. 16/D F Packed 1 set of 5 pieces in pasteboard box.
Suitable for 14 inch and 16 inch Dolls.......... Per set, $1.50
Size of table 6 inches high, 8 inches in diameter, chairs in proportion.

No. 19/D F Packed 1 set of 5 pieces in pasteboard box.
Suitable for 19 inch and 21 inch Dolls.......... Per set, $2.50
Size of table 6½ inches high, 10 inches in diameter, chairs in proportion.

18

Dolly Walker Dolls

While Harry Coleman, a ventriloquist from England, was performing in the United States, circa 1917, he invented a doll he called "Dolly Walker." Coleman received patents for the doll, which was made mostly of wood with hinged and spring joints.

According to the Colemans, writing in their book, *The Collector's Encyclopedia of Dolls*, Harry Coleman purchased parts for the dolls from Emma Clear, who owned the Humpty Dumpty Doll Hospital. These probably included the composition shoulder heads and lower arms that were regular doll parts.

By 1920, the Wood Toy Co. was advertising the dolls. They may have assembled the dolls from the beginning or joined the enterprise circa 1919-1920. An ad in *Playthings* of August 1920 stated that the dolls were 28" tall, came with painted or sleep eyes, and could be purchased with painted hair or real hair wigs.

The Colemans also state that the dolls came in sizes of 18", and 24" in 1921. If this information is correct, these dolls must be very rare. According to the authors, Emma Clear purchased a half interest in the patent rights for the Dolly Walker dolls circa 1922. She may have become more involved in their production at this time. There are some distinct differences in the bodies of the dolls, which could indicate that a change was made in the doll assembly at some point. Most of the early dolls were stamped on the back of the shoulder plate "This is a Dolly Walker/Patent Pending." They are made with cardboard bodies and a wood frame. Later dolls carry the stamp "Patented in the USA/Other Patents Pending/Patents Applied For in All Other Countries." These dolls have screen wire bodies and fuller legs. Perhaps the changes were made to fit another company's tools.

Information in the Coleman book indicates Harry Coleman died in 1923. No further information is known about the Dolly Walker after that time.

The two Dolly Walker doll bodies pictured here show distinct differences. The doll on the left has a body made of cardboard, while screen wire has been used for the Dolly Walker body on the right. Both dolls have composition shoulder heads, arms with hinges at the elbow, and composition lower arms. The more recent doll on the right has a stamped piece of wood under his body that reads "Patented in the U.S.A./Other Patents Pending/Patents Applied For in All Other Countries." Besides the different body coverings, the two dolls have some differences in their legs. What appears to be the older doll has thinner legs while the newer model has heavier legs. A change has also been made in the size and placement of the piece of wood between the legs. The cardboard body doll has the smaller wood and the wire screened body doll has the larger wood support under the body. A change in the design of the top of the leg under the spring has also been made. The fuller legs are cut clear to the back of the wood on each side while the cut in the thinner leg doll is only on the front, approximately halfway through the leg. Both legs have the same spring hinge joint. (See the "World War I Related Dolls" chapter for a different walking doll from the era.)

28" tall Dolly Walker doll, circa 1917-1919, pictured on the right. The doll was designed and patented by English ventriloquist Harry H. Coleman while he was performing in the United States. The doll has a composition shoulder head (with some repainting), lower composition arms, a cardboard type body over a wood frame, and wood legs and upper arms. She has painted brown eyes, an open-closed mouth with two teeth, and a real hair wig. Her knees are connected with a spring hinge and a regular spring attaches her legs to the body. Her elbows also have a hinge so they can bend. Most of these early dolls were stamped on the back of the shoulder plate "This is a Dolly Walker/Patent Pending." The doll on the left is assumed to be a later Dolly Walker doll. He has a composition shoulder head with sleep eyes, an open mouth with teeth, and a replaced wig and clothing. According to the Colemans, writing in *The Collector's Encyclopedia of Dolls*, Coleman purchased doll parts from Emma Clear (Humpty Dumpty Doll Hospital), and she purchased a half interest in the patent rights in 1922. Perhaps her doll hospital in California assembled some of the dolls. (Doll on right with repainting on her face \$150, doll on left \$175+).

Chapter Seven

Comic and Celebrity Dolls

Comic Dolls

The first real comic strips in the United States were created and printed in the country's newspapers in the late 1800s and early 1900s. They included "The Brownies," 1880s; "Alphonse and Gaston," 1902; "The Yellow Kid," 1895; "Buster Brown," 1902; "Foxy Grandpa," 1900; "Happy Hooligan," 1900; "The Katzenjammer Kids," 1897; "Little Nemo in Slumberland," 1905; "Mutt & Jeff," 1907; "The Newlyweds," 1904; and "Toonerville Folks," 1908.

Readers loved this new entertainment. Comic strips were popular before folks could attend movies or listen to the radio, and this love affair continued through the 1940s. Comic books, based on the most popular comic strip characters of the 1930s and 1940s, are still hot collector items today. In addition, the strips inspired many spin-off products, including toys and dolls.

In 1910, many of the early strips were still appearing in newspapers as a daily feature, on Sundays, or both. From 1910 to 1930, other famous strips were created. These included "Barney Google," 1919; "Blondie," 1930; "Bringing Up Father," 1913; "Ella Cinders," 1925; "Felix the Cat," 1923; "Gasoline Alley," 1918; "The Gumps," 1917; "Harold Teen," 1919; "Joe Palooka," 1928; "Krazy Kat," 1910; "Little Annie Rooney," 1929; "Little Orphan Annie," 1924; "Little Mary Mixup," circa 1919; "Moon Mullins," 1923; "Nize Baby," 1927; "Peter Rabbit," 1920; "Popeye," 1929; "Skippy," 1925; "Smitty," 1922; "Tarzan," 1929; "Tillie the Toiler," 1921; "Toots and Casper," 1919; and "Winnie Winkle," 1920.

Because the strips were so popular, the German Hertwig & Co. produced bisque nodder dolls in the images of characters from several of these comics during the late 1920s and early 1930s. Included were figures of characters from "Mutt & Jeff," "Little Orphan Annie," "Harold Teen," "Gasoline Alley," "Smitty," "Moon Mullins," "Winnie Winkle," and "The Gumps."

The Chicago Live Long Toys produced stuffed comic dolls made of an oilcloth type material during this same time period. Their dolls included those based on characters from "Gasoline Alley, "The Gumps," "Smitty," "Moon Mullins," "Harold Teen," "Winnie Winkle," and "Freckles and his Friends."

Some of the most collectible of the comic strip dolls are the wood dolls produced by the A. Schoenhut Co. The firm produced both Maggie and Jiggs from the "Bringing Up Father" strip plus Barney Google and Spark Plug from the "Barney Google" strip in doll form during the early to mid 1920s.

The Swiss A. Bucherer firm manufactured metal comic character dolls tied to the United States. They included "Mutt and Jeff," "Maggie and Jiggs," "The Katzenjammers," and "Happy Hooligan."

In addition, the Jaymar Specialty Co. of New York City also issued a series of wood comic strip related dolls in the late 1920s and early 1930s.These were jointed dolls with molded painted clothing. Included were dolls representing characters from "Little Orphan Annie," "Popeye," "Joe Palooka," "Krazy Kat," "Moon Mullins," and "Bringing Up Father."

The doll companies also manufactured cloth dolls based on the comics during this period. Little Orphan Annie was made of composition and cloth circa 1931 while Nize Baby (from the "Nize Baby" strip) and Buttercup ("Toots and Casper") were both manufactured of cloth in the late 1920s. Earlier cloth dolls had also been produced in the image of Foxy Grandpa, circa early 1900s.

E.I. Horsman was responsible for producing several comic connected dolls. These included Little Nemo in 1911 ("Little Nemo in Slumberland"), Gene Carr Kids, circa 1915 ("Lady Bountiful"), and Little Mary Mix-up from the strip of the same name in 1919. Later, in 1928, the Horsman Co. marketed another comic doll in the image of popular Ella Cinders.

Ideal Novelty & Toy Co. issued both Mr. Hooligan and Snookums dolls ("The Newlyweds") in 1910. The heads of both dolls were made in Germany, but perhaps Ideal was responsible for adding the Teddy Bear-like bodies. A collectible Buster Brown doll was manufactured by Ideal during the 1920s.

Bobby Blake and Dolly Drake, the comic characters designed by Grace Wiederseim Drayton, were produced in doll form by the Louis Amberg & Son firm, circa 1911. Dolly Dimple and Bobby Bounce, Drayton's later comic characters, were also marketed as dolls, circa 1931. These dolls were all-bisque and were made in Japan.

The Effanbee firm also manufactured a comic character doll during this period. He was based on the Skippy character, which began as a comic strip in 1925. The Effanbee composition doll was issued in 1929. An all-bisque Skippy was later produced in Japan. He had jointed arms and stationary legs.

The Kewpies, Brownies, and HEbee SHEbee characters have all been placed in this chapter because their popularity first developed in the print media as drawings in newspapers, magazines, and books. A variety of dolls from many different firms have been produced in these characters' images.

Barney Google

The "Barney Google" comic strip character first appeared in the *Chicago Herald Examiner* in 1919. The strip by Billy DeBeck involved the American sports scene. In the 1920s, Barney was involved with a race horse named Spark Plug and the strip became very popular. The strip moved on to feature other characters beginning in the mid 1930s.

7.5" tall wood Barney Google and his 6.5" tall horse Spark Plug, both made by the A. Schoenhut Co. circa 1922. They were based on the comic strip originated by Billy DeBeck in 1919. Barney appears to be all original wearing a wood top hat, white shirt, black felt coat, red tie, and red and white checked pants. He is jointed at the neck, shoulders, and hips. He has a large applied nose, painted and molded features, and large feet. He carries his original tag on his left foot. Two different Barney figures were produced. The other example has applied ears and wears a bow tie. Spark Plug also includes his paper tag. Both tags read "Copyright 1922 by King Features/Syndicate, Inc./Patent Applied For." Sparkplug is jointed at the neck and legs. He is missing his blanket and his ears and tail have been replaced. The set that includes Barney with the applied ears was advertised in the Butler Bros. catalog for 1925. The wholesale price was approximately $1.60 for each set. Pictured with Barney and Spark Plug is the "Barney Google" sheet music written by Billy Rose and Con Conrad with the permission of DeBeck, the originator of the comic strip. It carries a copyright of 1923 from Jerome H. Remick & Co. ($300 for set, sheet music $12.50).

Bobby Blake & Dolly Drake

The "Bobby Blake & Dolly Drake" comic characters were first drawn by Grace Gebbie Wiederseim (Drayton), circa 1907. The series was carried by the *Philadelphia Press*. A series of children's books was also marketed that featured the characters.

"DOLLY DRAKE" AND "BOBBY BLAKE."
48c and 98c

These original characters are well known to children. Genuine unbreakable composition heads, painted in fast colors. Safe in hands of smallest child. Stuffed silesia body, riveted hips and shoulders. Neatly made removable clothes in light colors. Large size dolls illustrated, 11½ inches high, with large heads and full bodies. Smaller dolls are of same type, but of smaller proportions and not so elaborately dressed.

No. 18G23271 "Dolly Drake," large size. Price......**98c**
No. 18G23272 "Bobby Blake," large size. Price......**98c**
Shipping weight, 1 pound.
No. 18G23273 "Dolly Drake," small size. Price.....**48c**
No. 18G23274 "Bobby Blake," small size. Price.....**48c**

Grace Drayton's Bobby Blake and Dolly Drake dolls were based on comic characters the artist created for the *Philadelphia Press* They were advertised in the Sears & Roebuck catalog in 1911. The dolls were produced by the Louis Amberg & Son New York firm in 1911. They were originally tagged listing G. G. Wiederseim as their originator (Wiederseim being Drayton's first husband's name). Both dolls had composition heads and cloth bodies and limbs. Their hair was molded and painted, as were their features. They came in two sizes in the advertisement. The largest size was 11.5" tall and the dolls sold for 98 cents each. The smaller size sold for 48 cents each. *Courtesy of Ronald S. Barlow from his book,* The Great American Antique Toy Bazaar 1879-1945.

Bringing Up Father

"Bringing Up Father" was created by George McManus and became a regular Hearst strip in 1916. Jiggs and Maggie were the two main characters, who moved from the lower class to the upper class by winning the Irish Sweepstakes. Jiggs still enjoyed his old friends at Dinty Moore's tavern while Maggie enjoyed her new place in high society. These two different life styles caused problems between the married couple.

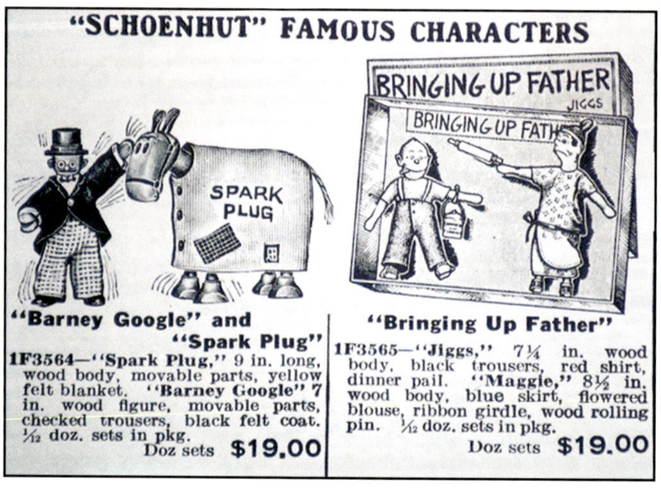

"SCHOENHUT" FAMOUS CHARACTERS

"Barney Google" and "Spark Plug"

1F3564—"Spark Plug," 9 in. long, wood body, movable parts, yellow felt blanket. **"Barney Google"** 7 in. wood figure, movable parts, checked trousers, black felt coat. 1/12 doz. sets in pkg.
Doz sets **$19.00**

"Bringing Up Father"

1F3565—"Jiggs," 7¼ in. wood body, black trousers, red shirt, dinner pail. **"Maggie,"** 8½ in. wood body, blue skirt, flowered blouse, ribbon girdle, wood rolling pin. 1/12 doz. sets in pkg.
Doz sets **$19.00**

Advertisement from the Butler Bros. catalog in 1925, showing Schoenhut's Barney Google and Spark Plug along with Maggie and Jiggs dolls from "Bringing Up Father." The ad was reprinted in the book *The Great American Antique Toy Bazaar 1879-1945*. Dover Publications now publishes the book. *Courtesy of the author, Ronald S. Barlow.*

9" tall Maggie and 7" tall Jiggs from the comic strip "Bringing Up Father" by George McManus. The figures were produced by the A. Schoenhut firm, circa 1925. They have molded hair, painted features, and large feet. Jiggs is smoking his trademark cigar. They are wearing their original clothes but Maggie originally came with a rolling pin and Jiggs carried a dinner bucket. ($600+ a pair). *Dolls and photograph from the collection of Susie's Museum of Childhood at Bluebird Farm in Carrollton, Ohio.*

Buster Brown

The cartoonist Richard F. Outcault began drawing "Buster Brown" cartoons in 1902. He was used in a Sunday strip for *The New York Herald*. Buster was supposed to be a ten-year-old boy who lived with his well-to-do family and his dog Tige. He continued in strip form until 1920. After that, the character was used mostly in advertising for Buster Brown Shoes. Several different dolls were produced in the character's image, including a small all-bisque doll that featured jointed arms and molded clothing.

12.5" tall Buster brown doll, circa early 1920s, with a composition head, molded, painted hair and features, and a cloth body and limbs. His arms are hooked to the body with outside discs. He appears to be all original including his hat. He may be missing a large bow at his neck. He is wearing a metal pinback that reads "Buster Brown Club/Buster Brown Bilt Shoes." Although Buster Brown was created as a comic character by Richard F. Outcault for *The New York Herald* in 1902, the character gained additional fame as the advertising character for Buster Brown Shoes. ($250-300). *Doll from the collection of Ellyn McCorkell.*

Brownies

The Brownies were created by Palmer Cox and began appearing during the late 1800s in various print materials. A variety of Brownie cloth dolls made by the Arnold Print Works was sold during the late 1800s and early 1900s. A series of twelve dolls produced in 1892 was particularly popular.

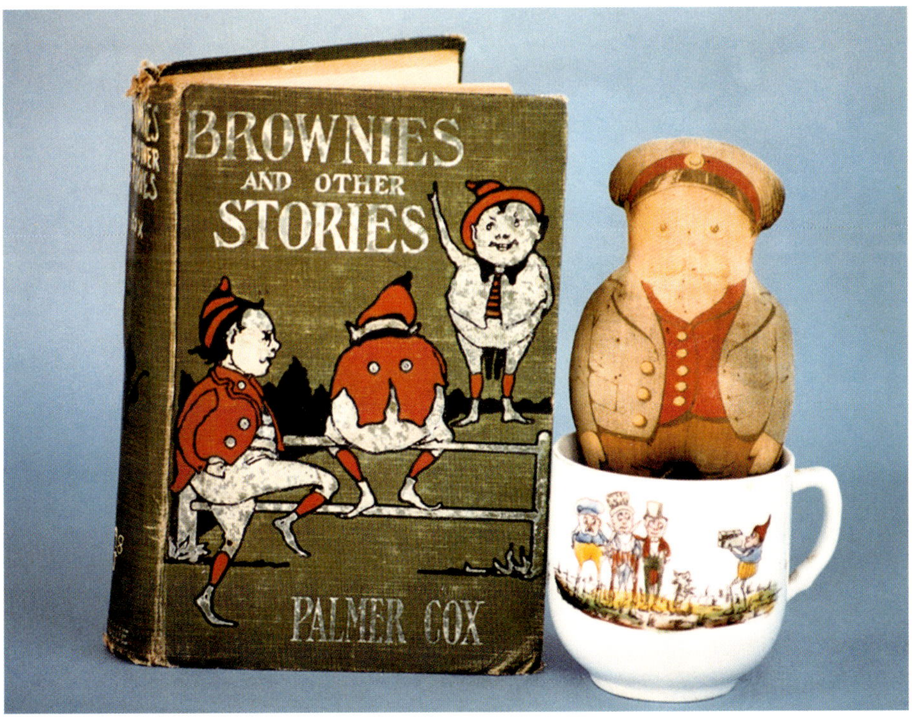

7" tall, all cloth Palmer Cox Brownie doll, circa late 1890s-early 1900s. He was probably made by the Arnold Print Works of North Adams, Massachusetts. The dolls were marked "Copyright 1892 by Palmer Cox." Twelve different Brownie dolls were included in a yard of material that sold for 20 cents, according to the Colemans writing in *The Collector's Encyclopedia of Dolls*. Brownies included the Chinaman, Dude, German, Highlander, Indian, Irishman, Soldier, Sailor, John Bull, Uncle Sam, Canadian, and Policeman. The pictured Brownie appears to represent John Bull. The dolls were sold uncut and were to be cut out, stitched, and stuffed at home. Pictured with the cloth doll is a Brownie decorated cup and an early book by Palmer Cox called *Brownies and Other Stories*. (Faded doll $40-50, cup $25, book with wear $20).

Dolly Dimples

The "Dolly Dimples" comic feature was first created by Grace Drayton in 1915. Bobby Bounce was also a character in the series.

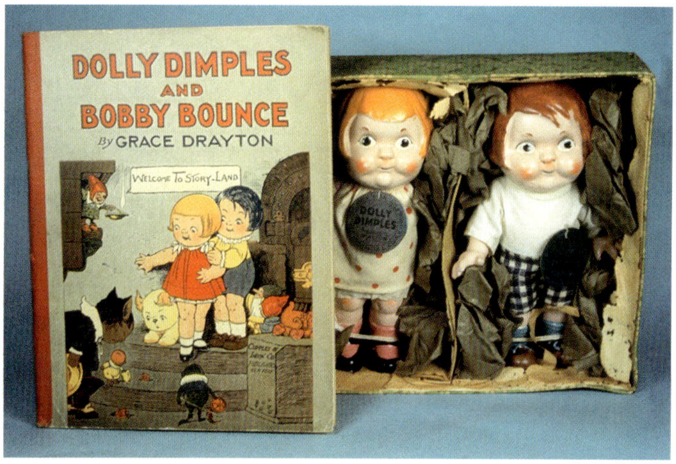

7" tall all-bisque Dolly Dimples and Bobby Bounce dolls, designed by Grace G. Drayton. The dolls are still tied in their original box. They have painted molded hair and features, movable arms, and stationary legs. Their shoes and socks are also molded and painted. The front of the box reads "Dolly Dimples/Bobby Bounce/by Grace G. Drayton/Pat. Ap. 1931/#83897/Sole Dist. Geo. Borgfeldt & Co. New York/Made in Japan/66.1545 1/6 doz." The dolls still have their original tags. Pictured with the dolls is a *Dolly Dimples and Bobby Bounce* book by Grace Drayton. It includes four squares of comic strips on each page and was published by Cupples and Leon Co., New York, in 1930. The book was based on the artist's comic feature "Dolly Dimples," which began in 1915. (Not enough examples to determine a price). *Dolls and book from the collection of Marge Meisinger. Photograph by Carol Stover.*

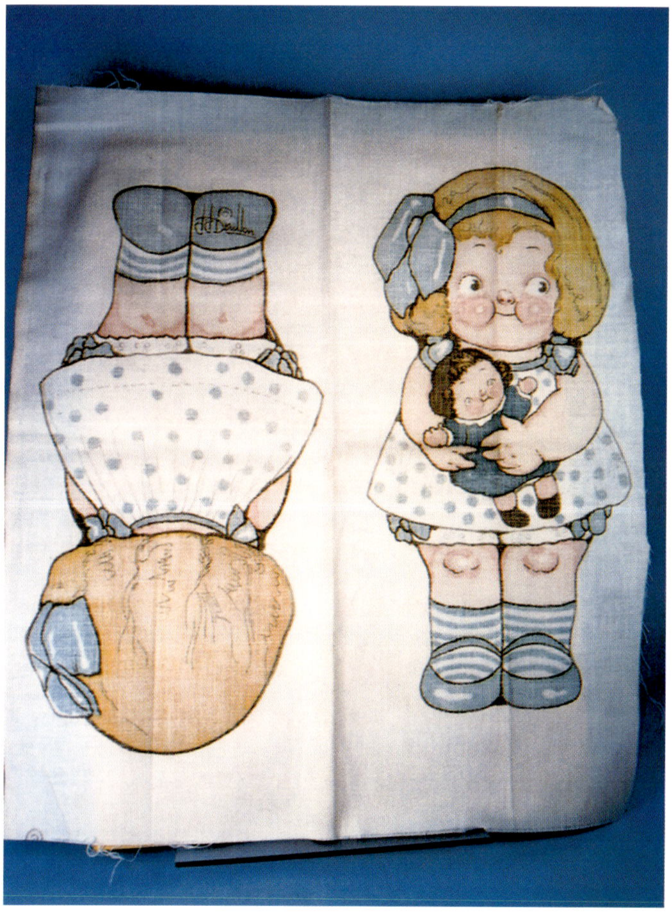

Ella Cinders

Bill Conselman and Charlie Plumb's "Ella Cinders" began as a comic strip in 1925. Its setting was Hollywood and the plot followed a poor Cinderella character called Ella Cinders as she rose to fame and fortune after securing a movie contract. The strip focused on her Hollywood adventures and her many romances during the 1920s. It lost popularity in later years, but remained a strip until the 1950s.

18" tall Ella Cinders composition doll made by the E.I. Horsman firm in 1928. The doll has a composition swivel head and jointed composition arms and legs. Her body is cloth. She has painted and molded hair and features. Ella wears her original tagged dress with replaced shoes and socks. She is marked "1925 MNS 18." The doll was based on the Ella Cinders comic strip by Bill Conselman and Charlie Plumb, which originated in 1925. (400+). *Courtesy of Frasher's Doll Auctions Inc., Oak Grove, Missouri.*

11.5" tall uncut fabric doll designed by Grace G. Drayton, circa 1920s. She appears to be Dolly Dingle, the famous character featured as paper dolls in many *Pictorial Review* magazines beginning in 1913 and continuing in many of the monthly magazines until 1933. This uncut cloth doll is marked on the foot "© G.G. Drayton. ($75+). *Doll from the collection of Marge Meisinger. Photograph by Carol Stover.*

Foxy Grandpa

The "Foxy Grandpa" comic strip was created by Charles Edward Schultze in 1900. The main character was an older man called Foxy Grandpa. Two young mischievous boys also appeared in the strip as they tried to get the best of Grandpa. The strip ended circa 1918.

20" tall cloth Foxy Grandpa thought to have been made by the Art Fabric Mills circa 1905. The doll came in two pieces and was to be stitched together and stuffed. He was based on the "Foxy Grandpa" comic strip by Charles Edward Schultze drawn for *The New York Herald*. It began on January 1, 1900 and ended in 1918. ($150-225). *Doll and photograph from the collection of Jean M. Grout.*

12" Foxy Grandpa and two 9" boy dolls based on characters from the comic strip. The dolls have stuffed stockinet heads, applied eyes and hair, and painted features. The bodies and limbs are cloth. These dolls are wearing their original outfits made of cloth and felt. The dolls were sold as a set. ($1,200-1,400 set). *Grout Collection.*

Gasoline Alley

"Gasoline Alley" began as a comic strip in 1918. Frank King, the creator of the strip, used the concept of aging his characters as time passed, which no other cartoonist had done. Walt Wallet, a bachelor, was the main character in the beginning, but in 1921 he found a baby boy abandoned on his doorstep. He adopted the child and named him Skeezix. Walt married Blossom in 1926 and they had two more children. The strip followed Skeezix as he grew up, served in World War II, married, and began his adult life.

12" tall Skeezix oilcloth type doll from the comic strip "Gasoline Alley" by Frank King, which began in 1918. The doll has painted hair, features, and clothing. His back is stamped "Skeezix King Features." He was made by Live Long Toys based in Chicago. A similar Skeezix was offered in the Sears 1927 catalog. It came in two sizes including a small 9.75" size priced at 39 cents and a larger 13.5" tall doll for 79 cents. Other "Gasoline Alley" characters made by Live Long Toys include Uncle Walt, Auntie Blossom, Rachel (black maid), and Corky (little brother). The German Hertwig & Co. firm also produced bisque nodders to represent Gasoline Alley characters. They included Skeezix, Uncle Walt, and Auntie Blossom. Pictured with Skeezix are two Skeezix books including *Skeezix Out West* by Frank King and *Skeezix Color Book* published by McLoughlin Brothers, Inc. 1929. (Skeezix doll $40 with some wear, books $25 each).

Gene Carr

Gene Carr created the "Lady Bountiful" comic strip for *The New York Herald* in 1904. The "Gene Carr Kids" dolls were based on characters from "Lady Bountiful," according to information on an original tag described in the book *Composition Dolls Vol. II 1909-1928* by Polly and Pam Judd.

14" tall Blink (also called Skinny), one of the dolls known as the Gene Carr Kids marketed by E.I. Horsman circa 1915. He has a composition head and hands and a cloth body and limbs. The doll has molded painted hair, painted features that include a big smile, and eyes molded shut. He is jointed at the shoulders and hips. The clothing may be original. The dolls were meant to represent comic characters drawn by Gene Carr for the *New York World* "Lady Bountiful" strip. Others in the series included Mike, Lizzy (Skinny dressed as a girl), Jane, and Snowball (black character later replaced by Smoke). ($300-350). *Doll from the collection of Ellyn McCorkell.*

Close-up of Blink's face, also used on the Skinny and Lizzy dolls. *McCorkell Collection.*

The Gumps

Sidney Smith's "The Gumps" first appeared in *The Chicago Tribune* in 1917. According to the *World Encyclopedia of Comics*, "The Gumps" became the most famous and "widely read comic strip in the 1920s." The story line revolved around the three main characters of the Gump family: Andy (father), Min (wife), and Chester (son). After Smith's death in 1935, the strip declined in popularity.

4" tall Andy Gump and 2.5" tall Chester Gump bisque nodders, circa 1920s. The figures were based on "The Gumps" comic strip created by Sidney Smith in 1917. These German Hertwig & Co. products have heads that move. The firm also manufactured similar figures representing Min Gump (Andy's wife) and their Maid Tilda. The figures are marked "Andy Gump/Germany" and "Chester Gump/Germany." They are pictured with *The Gump's Painting Book*, by Sidney Smith. It is also circa 1920s. The book shows pictures in both color and black and white so the child could color the picture to match the example shown in color. What appears to be the same dolls were advertised in *Playthings* magazine in January 1929 by the Marshall Field department store. Comic characters were available from Gasoline Ally, Little Orphan Annie, Moon Mullins, Smitty, and the Gumps. Live Long Toys of Chicago also marketed a Chester Gump oilcloth type cloth doll during this period. (Andy and Chester $75 pair, paint book $25+).

Happy Hooligan

Fred Opper's "Happy Hooligan" first appeared as a comic strip in Hearst's newspapers in 1900. Hooligan was an Irish tramp who wore a tin can for a hat. The strip ran in one form or another until 1932.

A variety of Happy Hooligan dolls and figures were produced during the comic strip's heyday. Ideal advertised a "Mr. Hooligan" doll in *Playthings* in 1910. He had a German made composition head with a plush Teddy Bear type body. Ideal may have added the body to the German head. The Colemans, writing in *The Collector's Encyclopedia of Dolls*, list a Happy Hooligan doll made by Hahn & Amberg also in 1910. Several other figures of the comic character were made of composition. Some had jointed arms and legs and others were nodders. The Happy Hooligan character was also represented in the line of metal dolls manufactured by the A. Bucherer firm based in Switzerland during this period.

6.5" tall papier mâché doll with small molded hat and hair and painted features. He is jointed at the shoulders and hips and has molded shoes. His clothes appear to be original. He is stamped "Germany" on his back. His small hat gives the doll the look of the Happy Hooligan comic character from the strip of the same name begun by Frederick Burr Opper in 1900. It is not known if he is based on that character. He is pictured with a postcard featuring "Happy Hooligan" that is marked "Copyright 1906, by *American Journal Examiner.*" (Doll $50, postcard $10).

HEbee SHEbee

The bisque HEbee SHEbee dolls were based on characters drawn by Charles Twelvetrees called the "Twelvetrees Kids." The drawings appeared regularly in the *Pictorial Review* magazines and other print outlets during the late teens and 1920s. Like Rose O'Neill and Grace Drayton's drawings, Charles Twelvetrees drew his characters in the "cherubic" manner that was popular at the time.

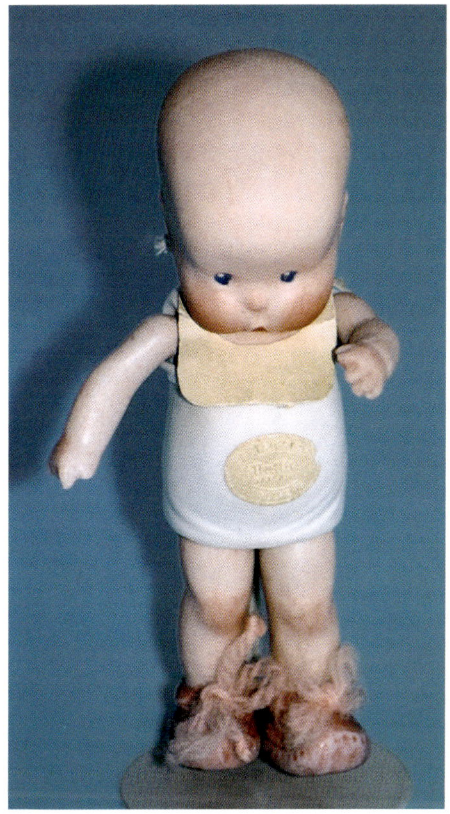

5" tall SHEbee all-bisque doll, circa 1926. The dolls were also dressed as boys and called HEbees. They were based on characters drawn by Charles Twelvetrees called the "Twelvetrees Kids." The dolls were also produced by E.I. Horsman in composition. Horsman was also responsible for the bisque dolls but they were made in Germany. The bisque dolls were jointed at the shoulders and hips, had painted hair and features, and molded clothing (chemise) and shoes. The shoelaces and what appears to be a bib were removable. This doll includes its original tag that reads "Copyright By/HEbee SHEbee/Trademark/Chas. Twelvetrees." ($650+). *Doll courtesy of Paula Reding and The Denver Doll Emporium.*

Kewpies

Rose O'Neill was the originator of the Kewpie figures circa 1909 when they first appeared in the *Ladies Home Journal*. By 1917, she was drawing Kewpies and verses for various newspapers. O'Neill also drew a panel featuring a Kewpie with a slogan or motto attached. She continued to expand her magazine drawings in the teens and 1920s and, in addition, the artist also designed many different Kewpie dolls that were issued from circa 1912 through the 1920s. The Kewpie dolls were first produced of all bisque but by 1915, both composition and celluloid Kewpies were for sale. In the 1920s, cloth Kewpies were also available. The bisque dolls were produced in Germany but many different companies from the United States manufactured the composition and cloth dolls during the different eras. One of the German firms involved in the bisque Kewpie production was the well known Kestner firm.

All-bisque Rose O'Neill Kewpie wedding dolls. The famous artist's first bisque Kewpies were produced circa 1912-13. The largest of the wedding dolls is 9" tall, the flower girls are 6" in height. The bride includes extra touches, including more eyelashes. The dolls have molded hair, painted features, and jointed arms. The groom has painted black feet for shoes. The dolls' clothing is all original. Information that came with the dolls stated that the set of dolls could be rented for weddings for $1.50 plus a $10 deposit. The costumes indicate that the dolls probably date from the 1920s. (Not enough examples to determine a price). *Courtesy of Paula Reding and the Denver Doll Emporium.*

12" tall, large all-bisque Rose O'Neill Kewpie. She has molded and painted hair and features with jointed arms and side glancing eyes. ($350+). *From the collection of Susie's Museum of Childhood at Bluebird Farm in Carrollton, Ohio.*

3.75" tall, all-bisque Rose O'Neill Kewpie "Huggers." The figures include no moving parts. ($150-200 pair). *Dolls from the collection of Ellyn McCorkell.*

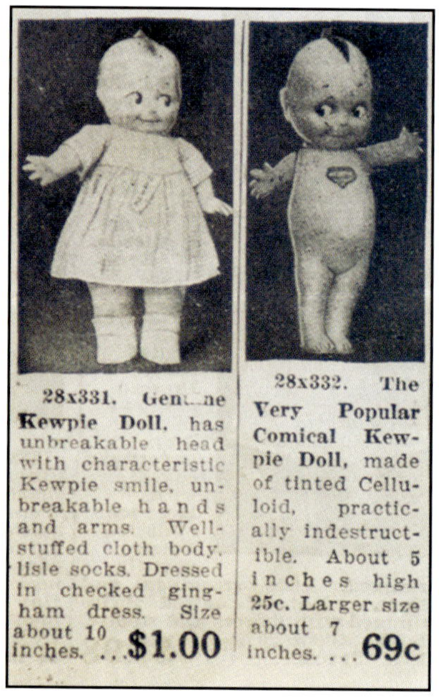

Advertisement for Rose O'Neill Kewpie dolls from the Bella Hess & Co. catalog, circa 1914-1915. The 10" tall Kewpie wearing a dress has a composition head and arms and a cloth body and legs. She is dressed in a checked dress that includes a cloth Kewpie tag at the waist. The doll sold for $1.00. The other Kewpie is made of celluloid and came in a 5" size for 25 cents and a 7" size priced at 69 cents. The doll included jointed arms. *Catalog from the collection of Marge Meisinger.*

9" tall twin "Cuddly Kewpie" dolls, tagged "Rose O'Neill." The dolls are covered with Jersey cloth, have painted flat faces, and are jointed at the shoulders and hips. They have tiny wings on their backs and a peak on top of the head. They are thought to have been made by Richard Krueger. ($200 each). *Dolls from the collection of Ellyn McCorkell.*

Cuddly Kewpie advertised in the Sears Fall and Winter catalog from 1931-32. The dolls came in sizes of 11.5" and 14" tall and sold for 98 cents and $1.49 each. These dolls were also covered with Jersey cloth.

14" all cloth Kewpie with a mask face and sateen body. She is all original. Her face looks exactly like the "Rabbit" Kewpie but she has no tag. She appears to date from the late 1920s to the early 1930s. ($300-400). *Courtesy of Paula Reding and the Denver Doll Emporium.*

12" tall (not including ears) Kewpie cloth "Rabbit." Perhaps this was a special Easter promotion doll. It includes its original tag that reads "Kewpie/Copyright Rose O'Neil/King Innovations, Inc. 119 W. 25th St. N.Y.C. Sole Distributors Made in U.S.A. Pat # 178580012." ($300-350). *Doll from the collection of Ellyn McCorkell.*

Little Orphan Annie

Harold Gray began the "Little Orphan Annie" comic strip in 1924 for the *Chicago Tribune – New York News Syndicate*. The strip remained very popular for decades. The main characters included Annie, Daddy Warbucks (millionaire), Punjab, and Annie's dog Sandy. Annie, Sandy, and Warbucks all were drawn with empty eye sockets. Many different dolls representing Annie have been manufactured over several decades from the 1920s through the 1980s, when the successful Broadway musical was turned into a movie.

10.5" tall Little Orphan Annie oilcloth stuffed doll based on the comic strip begun by Harold Gray in 1924. The back of the doll is stamped "Orphan Annie/Pat App For/Harold Gray." Her dog Sandy was also produced in the same form. These dolls were made by Live Long Toys in Chicago, Illinois, from 1923-1930. A similar Orphan Annie in a 13.5" size and wearing a coat and hat instead of a dress was priced at 79 cents in the Sears catalog for 1927. Her dog Sandy was 8.5" tall and also cost 79 cents. Other figures based on the "Little Orphan Annie" comic strip were also produced during the era. Included were the bisque nodder characters made in Germany by the Hertwig & Co. firm. Figures were made to represent Little Orphan Annie, her dog Sandy, and Daddy Warbucks from circa 1929 into the 1930s. The Jaymar Specialty Co. of New York City produced wood jointed Little Orphan Annie dolls in two sizes in the late 1920s or early 1930s. Pictured with the doll is *The Great Big Little Orphan Annie Paint & Crayon Book*, by Harold Gray. It was published in 1935 by McLoughlon Bros., Inc. (Doll $50-75, book in used condition $25).

14.5" tall Little Orphan Annie doll, circa 1931. The doll has a composition shoulder head and full arms, molded and painted hair, and painted features. Her body and legs are cloth. She is marked on the back "© FAMOUS/ARTISTS SYND." She is wearing her original underwear but has been redressed. She appears to be the doll advertised in the Sears Fall & Winter catalog for 1931-32. She sold for 45 cents and wore a red dress with white trim, was 14.5" tall, had a composition head and arms, and a cloth body and legs. Pictured with her is an electric Little Orphan Annie stove marked on the bottom "Metal Ware Corp. of Two Rivers Wisc." Little Orphan Annie and her dog Sandy are pictured on the back panel in silhouette. Under the picture is printed "Little Orphan Annie." The stove has two burners on the stove top and one in the oven. The oven door opens and a "pretend" wood oven regulator is mounted on the outside of the oven. The stove measures 12" tall x 13.5" wide x 6" deep, and is circa 1933-35. (Doll $50+ redressed and some wear, stove $125+).

Moon Mullins

The "Moon Mullins" comic strip was created by Frank Willard and first appeared in the *Chicago Tribune* in 1923. The main characters of the early strips included Moon Mullins (Moonshine), who lived in a rooming house run by Emmy Schmaltz. Emmy later married Lord Plushbottom. Moon's little brother Kayo also played a big role in the strip, as did Emmy's cook Mamie. The strip was most popular from the 1920s through the 1940s.

7.5" tall Moon Mullins and 4" high Kayo based on the" Moon Mullins" comic strip begun by Frank Willard in 1923. The figures were made by the Jaymar Specialty Co. based in New York City circa late 1920s-early 1930s. They are made of wood and are jointed at the neck, shoulders, and hips. The figures include molded clothing and painted features. Moon has a cigar in his mouth. The dolls are marked on the front of their bodies "Kayo/Willard" and "Moon/Mullins/Willard." Other dolls were also produced in the 1920s to represent Moon Mullins and Kayo. They include the oilcloth stuffed dolls in the image of Kayo and Moon Mullins made by the Live Long Chicago company and the bisque nodder figures marketed by the German Hertwig & Co. firm beginning in 1929. Emmy Schmaltz, who ran the important boarding house in the strip, was also one of the Hertwig's comic nodder figures. Pictured with the dolls is a *Moon Mullins and Kayo Big Little Book* published by Whitman Publishing Co. and marked "Copyright 1933, Frank Willard." (Dolls $100 each, book $25).

Max and Moritz

The German characters of Max and Moritz are included in this chapter as their creator Wilhelm Busch is considered by some to be the father of the comic strip form. His characters were in a German children's story circa 1865 that was told in pictures and text. Many dolls have been produced to represent these characters, most of which were made in Germany. Included are dolls made by Kämmer & Reinhardt and Recknagel. Dolls were made of all-bisque, as well as with bisque socket heads and a variety of composition bodies.

8" tall Moritz German character doll originated by Wilhelm Busch. He is marked "32 12/0." The doll has a bisque socket head with molded and painted hair, blue intaglio eyes, and a closed mouth. His body and limbs are composition. A companion doll called Max #32 was also produced. The famous Max and Moritz characters were made by several different companies through the years. This one was made by the German Recknagel firm. *Doll from the collection of Susie's Museum of Childhood at Bluebird Farm in Carrollton, Ohio.*

Mutt & Jeff

Bud Fisher's "Mutt & Jeff" comic strip had its beginning in 1907 and appeared under other titles until the Jeff character was added in 1908. By the 1920s, the strip, which featured tall thin Jeff and short stocky Mutt, was one of the most popular comics in the United States. The strip did not keep up with the times and was not as successful in later years.

Dolls were made of the two characters by the A. Bucherer Company in Switzerland and by Hertwig & Co., the German firm who produced all-bisque "nodder" dolls.

Right:
6.5" tall Mutt and 8" tall Jeff dolls made by the Switzerland firm of A. Bucherer, circa late 1920s. They were based on the characters from the "Mutt & Jeff" comic strip created by Bud Fisher in 1907. The dolls were advertised in the FAO Schwarz 1930 Christmas catalog and sold for $3.00 a pair. The heads, hands, and feet of the dolls are metal. The bodies include ball joints that enable the dolls to bend at the neck, shoulders, hips, knees, ankles, and wrists. Mutt appears to be all original, including his clothing, but Jeff is wearing replaced pants and originally wore a vest, shirt, and tie. ($500 a pair).

The Newlyweds

"The Newlyweds" was created by George McManus in 1904 for the *New York World*. The characters in the strip included Mr. and Mrs. Newlywed and their spoiled baby called Snookums. The baby had temper tantrums and the parents tried different ways to please the child. The *New York World* strip ended in 1912 when McManus joined the Hearst papers. There McManus established his more important strip called "Bringing Up Father."

Besides the German Snookums doll pictured, a 4.75" tall composition doll was also produced. It was jointed at the shoulders and hips and had painted molded clothing, hair, and features. Ideal advertised a Snookums doll in 1910, which used the same German composition head as pictured but was mounted on a plush body. A later Snookums doll was made by Madame Hendren in 1927. It was based on the movie, *The Newlyweds and Their Baby*.

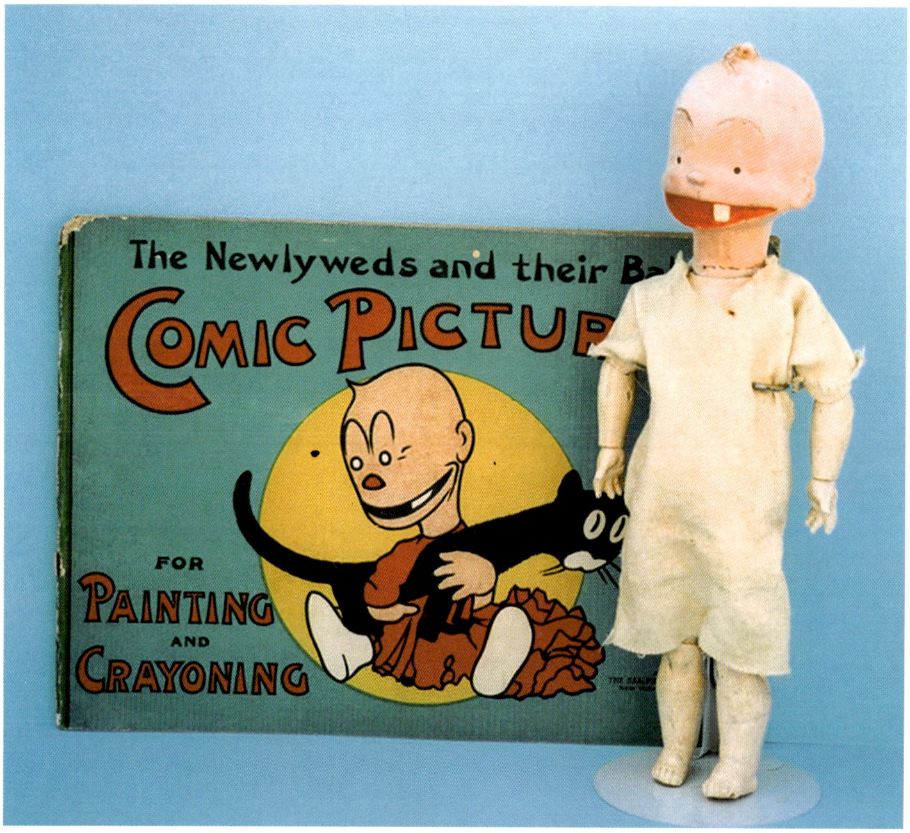

14.5" tall baby Snookums, circa 1910, from "The Newlyweds" comic strip created by George McManus for the *New York World* in 1904. His has a composition head with painted molded hair and features. His large open-closed mouth includes one tooth. He is on his original German ball-jointed body and is wearing what appears to be his original flannel clothing. It has been sewed together between his legs so perhaps it represents underwear. He is stamped on the back "Copyrighted/by George McManus/ Made in Germany." The doll is pictured with the book *The Newlyweds and Their Baby's Comic Pictures*, by George McManus. Published by the Saalfield Pub. Co. in Akron, Ohio in 1917, it is filled with both black and white and colored pictures from the popular comic strip. (Doll $300-350, book $50-75).

Nize Baby

The "Nize Baby" comic strip was created by Milt Gross beginning in 1927 for the *New York World*. It was based on characters from his earlier book of cartoons, also called *Nize Baby* and published in 1926. Gross specialized in both book and comic strips that featured Jewish immigrant characters. The Jewish dialect was written phonetically.

Advertisement for the cloth doll Nize Baby from the Sears 1928 catalog. The doll was under a heading called "Madame Hendren Novelty Dolls" in the 1927 Sears catalog. The cloth doll was 15.5" tall, had painted features, a baggy outfit, and felt shoes. His bib was marked "NIZE BABY" and he represented a character from the comic of the same name by Milt Gross. The doll sold for $1.27.

Skippy

"Skippy," drawn by Percy Crosby, became a comic strip character circa 1925. Skippy was a young boy who always wore a plaid hat and short pants. He was the leader of several other children who schemed against adults. The strip ended in 1942 when Cosby became ill.

5.5" tall, all-bisque Skippy doll, circa 1930s. He has molded clothes, hat, shoes and socks, and painted features. His arms are jointed. He is marked on the back "SKIPPY/© Percy L. Crosby/ Made in Japan." He is pictured with a *Skippy* Big Little Book #761 based on the "Skippy" Sunday newspaper comic by Percy L. Crosby. It carried a copyright from 1934. (Doll $50-60, Book, $20+).

Smitty

The "Smitty" comic strip was created by Walter Berndt as a daily feature in 1922. It was carried by the *Chicago Tribune-New York News Syndicate*. Smitty (Augustas Smith) and Herby (his younger brother) were key characters in the strip. During the 1920s, Smitty was an office boy who worked for Mr. Bailey.

Smitty and Herby were produced in doll form by several companies. The German Hertwig firm marketed both Skippy and Herby as bisque nodders during the late 1920s to the early 1930s. The Chicago based Live Long Toys also manufactured oilcloth type stuffed dolls of both Smitty and Herby during the same period. Unidentified cloth dolls were also produced of both boys circa 1930s. The dolls were 13.5" tall (Smitty) and 11.5" tall (Herby). Both characters included some removable clothing. Small all-bisque figures of the comic characters were "Made in Japan" during these same years. These dolls had no moving parts and included molded painted clothing.

Two 2.5" tall Herby all-bisque figures representing Herby who was Smitty's little brother in the "Smitty" comic strip created by Walter Berndt in 1922. The doll on the right has no moving parts. He is marked on the back "HERBY/Made in Japan." The other doll is a nodder and was made by the German Hertwig & Co. firm. He is pictured in the Theriault book about the company. He is marked "HERBY" on his back. Both dolls have painted and molded clothing. (Nodder $25+, Japan model $15+)

Toots and Casper

Jimmy Murphy was the creator of the "Toots and Casper" strip, which appeared in the *New York American* in 1919. The characters included a young married couple called Toots and Casper (no last name) and their baby son Buttercup. The strip remained popular through the 1930s.

15" all cloth Buttercup, the baby from the "Toots and Casper" comic strip by Jimmy Murphy. She has painted features and a squeaker in her cloth body. She is wearing her original rompers but is missing her shoes and socks. She is marked on the back of her head "Buttercup/By Permission of/Jimmy Murphy/Copyright by King Features Syndicate, Inc." The doll was advertised in the Sears catalog for 95 cents in 1925. ($40-50 with stains).

Winnie Winkle

The "Winnie Winkle" comic strip was created by Martin Branner and made its first appearance in 1920. The strip was distributed by the *News-Tribune Syndicate* and was first titled "Winnie Winkle, the Bread Winner." The plot included events from Winnie's work place as a stenographer for Barnaby Bibbs and her romantic adventures. Winnie was the "breadwinner" for her father and mother as well as for herself. In 1922, an adopted brother Perry was added to the mix of characters.

3.5" tall, all-bisque Winnie Winkle "nodder," circa 1920s. She was based on the main character in the "Winnie Winkle" comic strip created by Martin Branner in 1920.This figure is pictured in the *Hertwig & Co. Archives 1890-1937* book as being made by the German Hertwig & Co. firm in the 1920s. She is marked "Winnie/Winkle/By/Branner." The head of the figure moves. Live Long Toys also produced a comic figure based on the "Winnie Winkle" comic strip. This firm based their comic character on Perry Winkle, Winnie's younger brother. He was made of an oilcloth type material and included printed clothing and features.

Celebrity Dolls

The tradition of producing dolls in the image of famous personalities, particularly movie stars, began almost as soon as moving pictures became popular. The stage and historical events also provided famous personalities who were used as models for dolls manufactured from 1910-1930.

The earliest of these types of dolls appears to be the doll made by Louis Amberg & Son to represent the actress Rose Melville in her stage role as Sis Hopkins. Melville made a career out of playing this character beginning as early as 1900.

The 14" tall doll has a composition head with molded braids on her head. Her head also included holes for the placement of ribbons. The body, arms, and legs are cloth.

Silent film stars John Bunny and Charlie Chaplin were also marketed in doll form by Louis Amberg & Son during the teens. Both of the dolls are very collectible.

The 1920s brought even more movie related dolls to the marketplace. Included were the Jackie Coogan doll in 1921 and the Baby Peggy dolls in 1923. The Coogan doll was issued by the E.I. Horsman Co. while Louis Amberg & Son was responsible for the Baby Peggy dolls.

Betty Bronson, a teen-age movie star during the 1920s, was also honored in doll form in 1926. She had appeared in several silent movies including *Peter Pan* in 1924 and *Ben Hur* in 1926. The doll was manufactured by the Averill Manufacturing Co. It was a topsy-turvy model, 13" tall, featuring composition shoulder heads with mohair wigs and sleep eyes on either end of the doll. The doll portrayed Cinderella, with clothing based on poor Cinderella and Cinderella at the ball costumes. Bronson had played the part of Cinderella in the silent film *A Kiss For Cinderella* in 1925.

Other silent firm stars honored with doll replicas include comedian Harold Lloyd, Our Gang members, and the most famous silent star of all, Rudolph Valentino.

Valentino secured lasting fame by developing a perforated ulcer and general peritonitis that caused his death at the age of thirty-one in 1926. He had already become a great male movie star with high sex appeal for the ladies. His successful films included *The Sheik* (1921), *Blood & Sand* (1922), and *The Son of the Sheik* (1925) among others. The star died in New York City and so many people arrived for the open viewing of his body that a riot broke out, injuring many of those in attendance. Valentino's "Viewing" and funeral became just as memorable as his movies and added to his movie legend.

It seems fitting that the most memorable Rudolph Valentino doll was produced in his native Italy by the Lenci firm. The 29" tall doll was made circa 1928. The doll had a sculptured face, was made of felt, and included a brown mohair wig and painted brown eyes. The doll was dressed in beige felt pants, a short jacket, a red and beige striped coat, a burnoose on his head, and leather boots with tassels. This doll would rank as the most valuable of the personality dolls of the period.

The most interesting historical personality doll of the time period was the one based on the image of Charles Lindberg, made in 1928. Lindberg became one of America's most famous aviators when he flew alone nonstop over the Atlantic from New York to Paris in 1927. "Lucky Lindy" is sometimes called America's last hero.

A 27" tall doll called "Our Lindy" was produced by the Regal Doll Co. in 1928 to honor America's hero. The doll had a specially sculptured face designed by Ernesto Peruggi and was made with a composition shoulder head, molded hair, and painted features. He wore a khaki flying suit trimmed with an imitation fur collar, belt, long gloves, and laced up boots. His original tag read "Our Lindy/America's Pride/MFD by Regal Doll Mfg. Co./N.Y.C./Copyright Reg 83336." The back of the shoulder plate is marked "© Regal Doll Co./Sculp E. Peruggi/1928." The doll with these markings is pictured in John Axe's book *The Encyclopedia of Celebrity Dolls.*

Baby Peggy (Montgomery)

Baby Peggy (Montgomery) was the most famous little girl in movies during the 1920s. Instead of being a light curly-haired beauty like Shirley Temple, Baby Peggy was a dark-haired cutie with a "Buster Brown" haircut featuring straight bangs. She made her first silent film in 1920 at the age of eighteen months. This film and others that came later were inexpensive two-reelers. Later movies included *Captain January, Little Miss Hollywood*, and *Helen's Babies*. The small actress left the screen in the mid 1920s. She tried for a screen comeback in the 1930s but received only a few bit parts. She later changed her name to Diana and became Diana Cary through marriage.

The Louis Amberg & Son firm marketed several different models of Baby Peggy dolls during 1923 and 1924. The larger dolls were made with composition or bisque heads. The smaller all-bisque dolls featured molded painted hair to represent Peggy's famous hair style. These dolls had jointed arms and legs. They ranged in size from 3" to 5.5" tall.

20" tall Baby Peggy (Montgomery) doll manufactured by Louis Amberg & Son in 1923. She has a composition shoulder head and lower limbs, and a cloth body and upper limbs. Her hair is distinctly molded to represent the tiny movie star's real hair style. Her features are painted, with her mouth molded in a smile with painted teeth. The doll could also be purchased with a wig. Each doll came with a celluloid pin that read "Baby Peggy of the Movies." She has been redressed. ($700-800). *Doll courtesy of Marge Meisinger. Photograph by Carol Stover.*

This Baby Peggy doll ad appeared in the Montgomery Ward Fall catalog in 1924. The doll was 18" tall and included a bisque head, brown sleeping eyes, and a bobbed wig. Her ball-jointed body was made of composition. Dressed in underwear, shoes, and socks, the doll sold for $1.98. Fully clothed in a bloomer dress and hat, the price was $2.98. *Catalog from the collection of Marge Meisinger.*

19" tall Baby Peggy with bisque head. In addition to the bisque head dolls with composition ball-jointed bodies, Baby Peggy dolls were also made with kid bodies, as in this example. Her shoulder head is marked "19©24/LA&S NY/Germany/50." She has brown sleep eyes, a brown mohair wig, bisque lower arms, and a kid body. The doll was made by Armand Marseille in Germany for Louis Amberg & Son. ($1,200+). *Doll courtesy of Paula Reding and the Denver Doll Emporium.*

John Bunny

John Bunny came to movies from the stage, where he had worked for many years. He was born in 1863 and died in 1915, only a few years after transitioning to the new form of entertainment called motion pictures. He worked in a number of silent comedy films for Vitagraph beginning in 1910. His size (very heavyset) was used to gain laughs from the audience at his expense. His films include *Pickwick Papers*, *Pigs is Pigs*, and *Bunny's Dilemma*. He made more than two hundred short movie comedies during his short movie career.

A doll was made in his image by Louis Amberg & Son in 1914. The doll was 14" tall and included a composition head with painted, molded hair. The original label on his sailor suit read "John Bunny Doll/Copyright L.A.&S. 1914 Trademark Registered/Made Exclusively by Louis Amberg & Son. N.Y./With consent of John Bunny/the Famous Motion picture Hero of the Vitagraph Co." The tag information is from the John Axe book, *The Encyclopedia of Celebrity Dolls.*

John Bunny doll advertisement appearing in the Bella Hess & Co. catalog for 1914-1915. The doll included a composition head made with molded hair, and painted features in the likeness of the early comedic motion picture star. It was produced by the Louis Amberg & Son firm in 1914. This example was dressed in a shirt and colored trousers. He came in two sizes: 11", priced at 69 cents, and 14", priced at $1.39. *Catalog from the collection of Marge Meisinger.*

Charlie Chaplin

Charlie Chaplin's "Little Tramp" character was probably the best known screen character in the world. With no language barrier in silent film, people could enjoy the Chaplin humor no matter what language they spoke. Chaplin made his first film in 1914, a one-reeler called *Making a Living*. In his second film, *Kid Auto Races in Venice*, he put together a tramp costume with borrowed pieces and the "Little Tramp" was born. The characters became so popular that many products were made to tie-in to the tramp persona. A doll made by Louis Amberg & Son in 1915 offered a very good likeness of Chaplin as "The Little Tramp." Two of Chaplin's biggest hit films of the era were "The Kid" from 1921, with child star Jackie Coogan, and *The Gold Rush* in 1925.

Chaplin's other successful silent films were made after talking pictures had taken over the movie houses. Included were *City Lights* in 1931 and *Modern Times* in 1936. Another of his notable films was his first talking picture, *The Great Dictator*, made in 1940. It featured Chaplin in a dual role as an Adolf Hitler type dictator and a Jewish barber.

14" Charlie Chaplin doll produced by Louis Amberg & Son in 1914. He has a composition head and hands with painted features, molded hair, and a cloth body. His clothes appear to be original. The label on his sleeve reads "Charlie Chaplin/World's Greatest Comedian/Made Exclusively by Louis Amberg/and Son. N.Y./by Special Arrangement with Essanay Film Co." ($650-700). *Doll from the collection of Ellyn McCorkell.*

Jackie Coogan

Jackie Coogan had a lifetime show business career but is best remembered for his role as the child in Charlie Chaplin's film, *The Kid*. He appeared in his first film at the age of eighteen months when he played a toddler in Skinner's *Baby for Essanay*. Chaplin picked him to appear in *The Kid* when he saw him working in a vaudeville act with his parents. When the film was released in 1921, Jackie was an instant hit. Other 1920s films made by the child star included *Peck's Bad Boy* (1921), *Oliver Twist* (1922), *The Young King* (1923), and *Little Robinson Crusoe* (1924). In addition to making many more films through the next five decades, Coogan also gained additional fans and publicity when he played Uncle Fester Frump on the ABC television series *The Addams Family* from 1964-1966.

In the 1930s, Coogan took his parents to court to try to secure some of the four million dollars he had earned as a child. Most of it had been spent by his parents so he received only $126,307. The "Child Actors Bill" that protected child stars' earnings was passed as a result of Coogan's bad experience.

Jackie Coogan Horsman doll advertised in the Sears Roebuck & Co. Fall and Winter catalog in 1923. The doll was 13.5" tall and sold for $1.48.

13.5" tall Jackie Coogan ("The Kid") doll, as advertised in the 1923 Sears catalog. He was marketed by the E.I. Horsman Co. beginning in 1921. He has a composition head and hands with a cloth body. The hair is molded and his features are painted. The doll is marked on the back of his neck "E.I.H.CO. 19©21." He is dressed in his original clothing and still has his "Jackie Coogan Kid" button. Also pictured is a metal pencil box from the period that features Jackie's picture in his costume from the Charlie Chaplin film *The Kid*, released in 1921. (Doll with paint loss $250-300, pencil box $50+).

Harold Lloyd

Harold Lloyd achieved fame as a comedy star of silent films. His most famous films included scary and funny stunts performed by Lloyd, usually in high places. Lloyd was born in Nebraska in 1894 and began his film career circa 1915, working for Hal Roach in *Just Nuts*. The fully developed character with the horn-rimmed glasses that he made famous appeared in 1921 in *A Sailor Made Man*. Other films showcasing this character included *Grandma's Boy* (1922), *Safety Last* (1923), and *The Freshman* (1925). His popularity declined with the coming of sound for movies. He died in 1971. His famous character is remembered with a cloth doll to be made from printed cloth.

12" tall Harold Lloyd cloth doll, circa late 1920s. Two dolls are pictured, one in uncut condition and the other in an assembled used condition. The doll was made in the image of Lloyd's movie persona. His most famous film character wore horn-rimmed glasses and a straw hat. His initials "H.L." are on his belt. The back of the doll includes a printed inscription reading "Yours for Happiness/Harold Lloyd." The doll was to be cut out, assembled, and stuffed after purchase. (Uncut $75+, Worn $35+).

Our Gang

The *Our Gang* comedies were begun by Hal Roach in 1922. All of the *Our Gang* films were shorts lasting thirty minutes or less. The early films from the 1920s were silent pictures. *Our Gang* films continued into the 1930s and 1940s with different cast members and, of course, as "talkies." The cast members from the early films included Mary Korman as Mary, Allen Clayton Hoskins as Farina, Jackie Condon as Jackie, Mickey Daniels as Freckles, and Joe Cobb as Fatty. The dog Spot also played an important role in these films. Mickey Daniels and Mary Korman also starred in another series of Hal Roach comedies called *The Boy Friend* in the early 1930s. Both also made feature films during the 1930s. Allen Clayton Hoskins appeared in feature films until he was fifteen.

Bisque "Our Gang" nodders were produced in Germany circa late 1920s-early 1930s. The dolls ranged in size from 2" (Spot) to 3.5" tall. They included five of the gang (one girl) and Spot. Dancing dolls were produced by Sayco representing all five of these early "Our Gang" members in 1926.

Advertisement for Our Gang Dancing Dolls from a Sears Special catalog in 1927. The dolls sold for 69 cents each. They were 12" tall and included composition heads and arms. The Mary doll was fitted with a bobbed style wig and Fatty and Farina had painted hair. The dolls danced when their mechanism was wound. Another ad featured in *Playthings* magazine for April 1926 pictured five dolls in a boxed set of "Our Gang Dancing Comedy Dolls." The dolls included Allen Clayton Hoskins as Farina, Jackie David as Jackie, Mary Korhman as Mary, Mickey Daniels as Freckles, and Joe Cobb as Fatty. The dolls were made by Sayco (Schoen and Yondorf Co. Inc.) in New York. *Catalog from the collection of Marge Meisinger.*

Chapter Eight

Boudoir and Pincushion (Half) Dolls

Boudoir Dolls

Boudoir dolls first became popular during the mid 1920s. These dolls were not meant as toys for little girls but were to be displayed in the bedrooms of adult women. Most of the early dolls were made mainly of cloth with either a cloth or composition head. The dolls resembled the 1920s female flapper with extra long arms and legs. Many of them were sold with no clothing or with just an added jacket or short skirt. Some were dressed by the women who purchased them. The cloth faces of the commercial dolls were either stiffened with starch or had shaped cloth faces of buckram. The bodies were made of sateen, velvet, or brocade. The American Stuffed Novelty Co. and the Unique Novelty Doll Co., both of New York, sold dolls of this type.

More expensive cloth boudoir dolls were produced by the English Dean's Rag Book Co., Norah Wellings in England, and the Lenci firm in Italy. These dolls were usually made of felt or velvet and were dressed in very fa ncy clothing made of organdy and felt. Most were made in the late 1920s and 1930s.

Many boudoir dolls of the period were produced with composition heads, and sometimes lower arms and legs. The Gerling Toy Co. of New York sold these types of dolls during the period.

The Montgomery Ward catalog for 1926 offered three different boudoir dolls for sale. A 30" example with a composition head and long cloth limbs and body, marketed as "to be dressed," sold for only 87 cents. She had painted features and a mohair wig. A 25.5" dressed model was priced at $1.98. The most expensive doll was 29" tall and wore a gown trimmed with marabou. She had hair of rayon floss and cost $2.59. The bodies of all three dolls were made of sateen.

The most sought after composition boudoir dolls are those made of all composition offered by the Mutual Novelty Co. of New York City. These dolls were heavily advertised in the trade magazines from 1925-1927. They were jointed at the neck, shoulders, elbows, hips, and knees. The most popular model is the doll with a cigarette in her mouth. In 1926, the firm advertised an Egyptian styled doll dressed in silk pajamas. One of the company ads stated that the dolls could also be purchased undressed.

Cloth boudoir dolls smoking a cigarette are also much in demand by collectors. *See Dolls and Accessories of the 1930s and 1940s* for a photograph of one of these dolls.

The more unusual boudoir dolls have become even more popular during the last several years. Besides the "smoker" and the "Art" dolls from Lenci, Norah Wellings, and other firms who made dolls of this type, collectors are looking for dolls in all mint or near mint condition, particularly those with floss hair, and wearing their original clothing and high heel shoes.

This ad appeared in the Butler Brothers catalog in 1929. It featured a variety of "Boudoir" dolls from the period. Included were inexpensive 5.5" tall dolls with china heads, and 10.5" and 16.5" tall dolls with composition heads. The advertisement also featured "cigarette" girls complete with cigarettes in a 21" size. All three sizes of composition head dolls had mohair wigs, painted features, and cloth bodies. The cigarette girls came with either velvet, wood, or composition slippers. Their costumes included a variety of designs of silk pajamas. *Ad from the collection of Marge Meisinger.*

In addition to those made with composition heads, many of the boudoir dolls of the period featured cloth heads. The pictured doll is 22" tall and has a molded cloth face and head, a mohair wig, and painted features. She has jointed arms and includes a cloth body and legs. She wears her original tattered dress with missing shoes, which may have been as simple as ribbons wound around her feet in a pleasing manner. ($75-100).

29" tall, all cloth boudoir doll thought to have been made to represent Josephine Baker, the famous American black entertainer who made such a hit in France. She has a molded face with painted features, and a brown body and limbs. The doll has no real hands or feet as these parts of her body have been covered with "grass" to match her original costume. (Not enough examples to determine a price). *McCorkell Collection.*

31" tall "put together" boudoir doll apparently made using pieces from several dolls. She is pictured here to show what an energetic collector can do to bring one of these early dolls back to life. Many of the advertisements of the period pictured a cloth boudoir doll that could be tied in a knot. Some had cloth bodies made of a designed brocade material as this doll does. Instead of cloth legs, she has very long composition lower legs measuring 8.5" that end in old-fashioned gold "made on" high-heeled shoes. Her lower arms and hands are also composition. Her excellent cloth head features a black mohair wig and molded and painted features. It may have been old store stock that has been added to pieces from an old doll. She has been dressed in a beautifully made outfit that reflects the styles of the 1920s dolls. Apparently the embroidery work on the jacket was cut from an article of adult clothing as was the matching satin material. Most collectors prefer original dolls but this doll with her long composition legs that reach nearly to her knees and her mint head offers an interesting new twist to the popular boudoir dolls. ($125-150).

23" tall cloth boudoir doll with a cigarette dangling from her mouth. She has a mohair wig, molded and painted features (with lots of eye shadow), and a red velvet body with her original matching jacket. Her high heel shoes also appear to be original. ($300). *Doll from the collection of Ellyn McCorkell.*

29" tall original boudoir doll, circa late 1920s. This flapper has a molded mask face with painted hair and features, cloth body and limbs, and composition hands. Her legs were made with material that matched her dress and head covering. Although the material on her legs has faded with time, her dress remains fresh. She wears her original high heeled shoes. A similar doll has been seen with a cigarette in its mouth. An embroidered pillow cover picturing a 1920s flapper is shown behind her. (With faded legs and missing one heel $85-100).

Right:
16" flapper doll from the 1920s made by Joanne Webster, an assistant to sculptor Larendo Taft. She has a painted and molded composition head, cloth body, silk and woven metallic fabric costume, and is unmarked. She is a one-of-a-kind doll. (Not enough examples to determine a price). *Doll from the collection of Marge Meisinger. Photograph by Carol Stover.*

25" tall jointed composition doll, sometimes called "The Parisienne, French Flapper, or the Cigarette Girl." The Mutual Novelty Corp. of New York City stated that they held the patent for this doll in an ad that appeared in *Toys and Novelties* in April 1925. The patent had been granted on April 14, 1925. The Konroe Merchants of New York were advertising the same doll in *Playthings* in November 1924. The doll is jointed at the neck, knees, elbows, shoulders, and hips. She has a real hair wig, painted features, and is smoking a cigarette. Her clothes (except for the scarf) are all original and appear on the dolls in both ads. She originally had a belt made of the same material as her jacket. This boudoir doll is one of the most popular of all dolls of this type. (Not enough examples to determine a price). *From the collection of Ellyn McCorkell.*

Boudoir dolls appear in almost all Lenci catalogs and have become a staple of their line. The male is model R-15 from the 1930-31 catalog and is 25" tall. The girl is from the 165 series and measures 26.5" tall. Male boudoir dolls are extremely rare. This is the only model known to be found thus far. (Male $2,000-2,500, female $1,500-2,000). *Dolls from the collection of Nancy Lazenby.*

Pincushion (Half) Dolls

Pincushion or half dolls, as they are now often called, are becoming more in demand. With the popularity of the Internet, treasures that a collector might never have seen in years of collecting can be purchased with the touch of a key.

Collectors are learning more about the various firms that produced these dolls and are more able to recognize the dolls made by such firms. Susan Endo describes the markings on the quality dolls made by the German Dressel & Kister firm in the December 1985 *Doll Reader*. According to Endo, the firms' half dolls are so tapered at the bottom they are not able to stand alone. This tapered red part of the doll is usually unglazed. Most of the dolls are marked with what looks similar to a child's drawing of half of the sun on its side connected to a longer straight line at the end. This firm's half dolls are beautifully made and most of the figures represent either Medieval times, the Colonial years, or the flappers and Pierrots of the 1920s.

Many of the half dolls produced by Hertwig & Co. of Katzhatte, Germany, can be identified by using the *Hertwig & Co. Archives* book by Florence Theriault. Although most of their half dolls are not as fine as those produced by Dressel & Kister, their designs were sold for many years and every collector will find several of their half dolls pictured in the book.

The half dolls issued by the German firm of F. & W. Goebel are also much in demand by half doll collectors. According to Sally Van Luven and Susan Graham, writing in their *Half Dolls Price Guide* book, the dolls were produced from around the end of World War I until the late 1930s. Some carry the Goebel trademark, which is a bee-shaped crown atop an intertwined "W and G". This firm represented famous personalities in a series of their dolls. These include Jenny Lind, Josephine (Napoleon's wife), and Marie Antoinette, as well as others. The firm also represented Colonial and Victorian ladies in their half dolls as well as dolls with wigs. This company reproduced several of its original half dolls in the 1990s.

The *Half Dolls Price Guide* also pictures a two page spread from a German Weiss, Kühnert & Co. catalog. It shows approximately fifty-five different designs of half dolls made by the company. The model numbers are shown under each doll. The dolls ranged in size from 2.75" to 4.5" tall.

According to authors Van Luven and Graham, one of the largest of the manufacturers of half dolls was the German Karl Schneider firm. Their mold numbers range from 10,000 to 20,000. This firm made nude, Colonial, flapper, and nicely dressed (molded) dolls. One of the most popular of the company's dolls is the Spanish lady sometimes called Carmen by collectors. It is model #10016.

Many more German firms produced half dolls as well as other companies around the world. Japan was particularly active in this production for several decades but these products are not in much demand. Many of the designs were based on German models but the Japanese products are inferior to the German half dolls and their prices remain low. The Fulper Pottery Co., located in Flemington, New Jersey, produced half dolls circa 1920. Although these dolls are not as well done as most of the German dolls, they are desirable because of their connection to the Fulper Pottery Co.

Probably the most important dolls in a half doll collection are those called the chocolate ladies. These half dolls include a doll with molded clothing who is carrying a tray that includes dishes. Some ladies carry a tea pot and cup and saucer, others carry only a cup and saucer, and the model made by F. & W. Goebel carries a pot, creamer, and sugar. These dolls also carry very high prices and usually cost $1,000 or more depending on the size and sculpturing of the figure.

The half dolls were used to decorate pincushions, brooms, dresser lamps, and bed lamps. Some collectors also include full figures or just heads in their collections. Other suitable additions to a half doll collection include perfume bottle figures and powder or powder puff holders made in figure form.

Linens from the 1920s and 1930s can also be added to a bedroom used to display a collection of half dolls. Included are embroidered bedspreads, pillow cases, sheets, fancy pillows, and dresser and vanity scarves. Boudoir dolls can decorate the bed and half dolls can be displayed on a dresser, bedside table, and wall shelves. The room, itself can be used as a guest room so the removal of the dolls will only need to happen infrequently.

Boudoir dolls and half dolls appear to be still gaining in value, and for a collector who wants to try something new, this field has plenty of items available in a wide variety of styles and prices.

Two large pincushion dolls made in Germany. The one holding a rose in her hand measures 4.5" tall. She is marked "Made in Germany." Her arms are away from her body and her decorations include a long molded necklace like those popular in the 1920s. The other doll is 4" tall. She wears a green hat and carries a fan. She is model number 7946 and she was produced by Hertwig & Co., the German firm located in Katzhutte. The company was in business from 1864 until it closed in 1990. The cloth pincushions have been added. (4.5" $200, 4" $100+).

Two German half dolls that include their arms away from the body. The china child doll measures 2.5" tall. She is holding a long flower arrangement. She is very finely molded and includes an unusual hair style. The other doll features a tiny hole in each hand so she can hold her skirt. She wears a molded bracelet on her wrist and the molding of the straps on her dress is also unusual. This half doll measures approximately 3" tall. The pincushions have been added. ($100+ each).

Half dolls were also made with wigs. Here are two German examples of these types of dolls. The one in the pink skirt was made by the German Weiss Kühnert & Co. firm. The half doll is 3" tall and includes her original wig. The ribbon and rose decoration in her hair appear to be original. Her pincushion skirt has been replaced. A strapless white top is molded on the front of the doll and both hands are molded to her body. The other wigged half doll is made of bisque. Several different firms produced similar dolls, including Herwig & Co. Some of the dolls had stationary arms while others included arms that moved, as does this example. Wigs came in both dark and light colors. This doll has side glancing eyes, and a turned head. She measures approximately 3" tall. She was molded nude. These types of dolls were popular in the mid to late 1920s. (china $75+, bisque $50-75).

Two small half dolls include a china one measuring 2.5" tall produced by the German Weiss Kuhnert & Co. firm. This model #5716 wears a bonnet with a bow on top and a white dress top. The doll also came in 2" and 3" sizes and with a blue bonnet and trim. Her hands are on her hips. The smaller unidentified German doll represents a child. She measures 2" tall, has long hair, and is wearing a plain white simple dress top. One of her arms is attached to the body and the other is partly away from the body. A half doll modeled in the image of a child is a desirable collectible because not many examples were produced. Pincushions added. ($50 each).

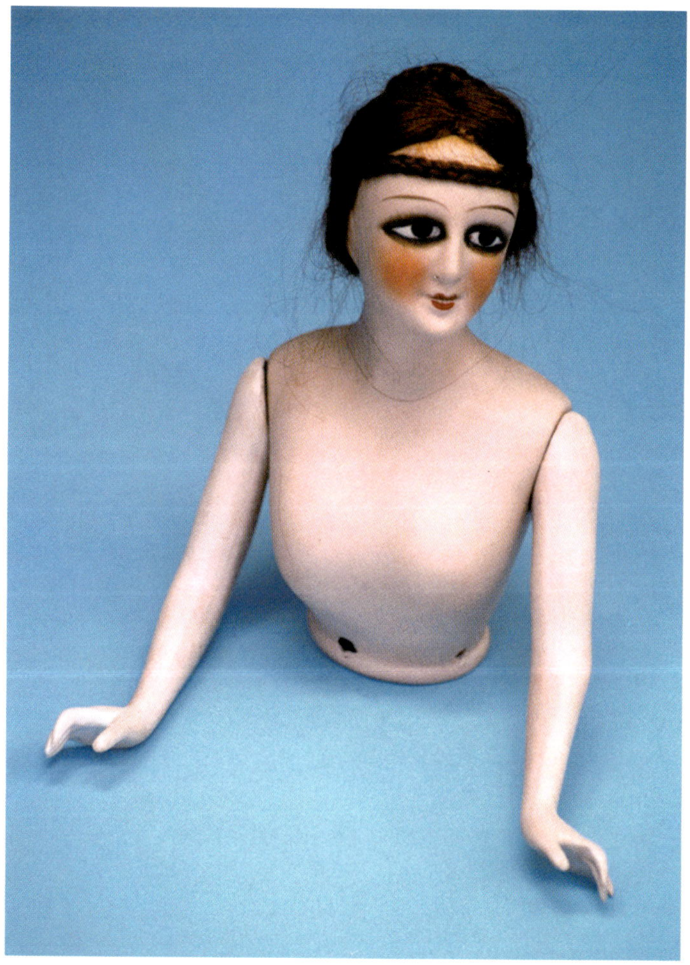

6.5" tall bisque half doll with movable arms. The doll includes large eyes accented with eye shadow, a closed mouth, and a real hair wig. Her head is turned to the side. A similar bisque doll with white complexion and a blonde wig has also been seen. The maker is unknown. ($350+) *Doll from the collection of Ellyn McCorkell.*

Colonial pincushion ladies became even more popular in the early 1930s as the bicentennial celebration of George Washington's birth (celebrated in 1932) drew near. Pictured are two different styles of German china half dolls of this type. The 3" doll wearing the green skirt was made by the Hertwig firm and is #6347. The doll came in several different colors. The larger doll is marked "Germany/502?" and is 4" tall Her china legs are marked 7983. Some of the china pincushion dolls were fitted with legs that were placed on the side of the pincushion. Both of the pincushions are replacements. (Hertwig $35+, Doll with legs $60+).

5.5" tall nude half doll with both arms away from the body. One hand is holding a rose. The doll has a colonial hair style and is wearing a black molded necklace. It is thought that she was made by the German F. & W. Goebel firm. Goebel also reproduced six half dolls in a tea cozy series circa 1985. They originally sold for $180.00 each, but seldom bring more that $75-100 on the current resale market. They were based on the firm's original designs. ($250). *McCorkell Collection.*

The nine dolls pictured here reflect the variety of half dolls that are available for today's collector. Top row: 4.5" #12756 open and returning arms with a flower, nude $85+; 5.5" Karl Schneider #15542 one arm extended, one open and holding a flower, $250; 4.5" chocolate lady holding a tray, $700. Middle row: 5.5" chocolate lady #9193 holding a tray, $800; 5.25" Dressel & Kister with arms extended holding a fan and with an elaborate beaded gold headdress and necklace, nude, $500+; 4.5" #1081 bisque half doll, nude with jointed arms and blue paperweight set eyes (not enough examples to determine a price). Bottom row: 6" Karl Schneider Carmen #10016, one arm extended, open and returning arm, molded lace mantilla, $500-900; 5.5" Volkstedt Rudolstadt, arms extended, large hat with plumes, nude, $500+; 6" Fulper Pottery, open and returning arms, $100. All prices are for mint dolls. *Photograph from Cobb's Doll Auctions, Johnstown, Ohio.*

Besides being attached to pincushions and lamp shades, half dolls were also used as tops for brushes in the 1920s. Pictured are two of these type dolls. The largest measures 8.25" tall including the brush. The brush is attached to a German half doll made by Hertwig & Co. The half doll itself measures approximately 3" tall. Her original clothing is intact and she also retains a gold tape hanger in the back. She is a very common doll with her arms attached to the body and a Spanish comb in her hair. The smaller brush doll is 7" tall. The half doll itself is only 2.5" tall. She also includes her abbreviated clothing, which covers her original marking if any is present. She may be a Japanese product as her modeling is not as nicely done as that on most of the German dolls. She has a pink painted hair band in her hair. The third item pictured is a more unusual pincushion that includes a china half doll at the top, separate legs mounted on a stand at the bottom, and a rounded pincushion in the middle. Even though she is stamped on her back "Japan," these pincushions are very collectible. She is 6" tall. (German broom $60, Japanese broom $30, Japanese pincushion $50+).

China half dolls were also used as lampshades in the 1920s. The smaller half doll is 3" tall. Including the frame, she measures 7.5" tall. She was made by the German Hertwig firm and is #5160. She is wearing a nicely decorated molded yellow cap, a molded tie at her neck, and a simple white top. Her arms are molded to her body at the front and back. She is still wearing her original shredded clothing. The frame under her body includes a metal piece to fit over a light bulb. It would appear to be a flammable shade if it were ever used for that purpose. The other half doll is pictured with its own skeletal shade. The 1920s half doll measures 4" tall. With the frame she is 8.5" high. She was made by the German Karl Schneider firm, which was one of the largest producers of half dolls. Her mold number is 14506. She is a flapper with a cloche hat and a coat that appears to have a molded ermine collar. One hand holds a purse to her body and the other one is perhaps holding her hat on in a strong wind. An electric porcelain bulb holder and a cord were originally connected to the frame. (Small with frame $65, large with frame $100).

Lamps and pin cushions that included half dolls made of a composition-chalk type material were also produced during this period. The lamps featured a half doll with arms away from the body and the pincushions were made with the arms molded to the body. The pictured lamp is 16" tall and its half doll has a wig and painted features. Her arms are attached to the body with wire. She has been redressed in vintage material. The largest half doll is 5" tall (11" with pincushion) and has also been redressed. Her head is similar to the one used on the lamp featuring a wig and painted features. The smaller pincushion half doll measures 4" tall (9" including the pincushion). She is all original. This half doll is also made of the composition-chalk material. (Lamp $75-100, pincushions $15+ each).

Besides displaying boudoir dolls and half dolls, no woman's bedroom in the late 1920s and early 1930s would have been complete without linens embroidered to match the lamp, dolls, and pincushions in the room. This ad appeared in the 1929-30 Fall & Winter catalog for M.W. Savage Co. of Minneapolis, Minnesota. It offers a matching bedspread, curtain drapes, 3-piece vanity set, pillow, and dresser scarf priced from 15 cents to 98 cents each. Of course, the catch was that these were stamped goods and each item had to be embroidered and finished in lace in order for it to be complete. In the late 1920s and early 1930s, however, more women stayed at home than were involved in the workforce. Many of these wives and mothers did embroidery or crochet projects as hobbies and were delighted to be able to improve their home décor using their sewing skills. The author's mother embroidered a bedspread using a design much like the one pictured. It was used as a summer bedspread until the colors of the thread had nearly disappeared. *Ad from the collection of Marge Meisinger.*

Examples of women's sewing projects from the late 1920s and early 1930s. Included are a dresser scarf and two pillows. A china trinket box to match the style of the period is also pictured. These items were used to decorate bedrooms. Some were pre-stamped and others were made using patterns sold in women's magazines or the "dime store." (Pillows $25 each, scarf $20, china lady trinket holder $35).

Chapter Nine

World War I Related Dolls

Doll production and marketing in the United States was greatly affected in many ways during the years surrounding World War I. Germany had been a primary source of dolls that were sold in America for decades. That country exported dolls made of china, bisque, metal, and celluloid, all of which were popular in the United States. Since this source of dolls was cut off for many years, new doll companies in the United States were founded and old firms were strengthened.

The fledgling Ideal, Effanbee, and Horsman firms increased their marketing and manufacture of composition dolls. Included in their new models were dolls made in the images of U.S. soldiers as well as Red Cross nurses. Although many different companies produced these types of dolls, probably the most popular is the Liberty Boy marketed by Ideal circa 1918. He had molded clothes instead of the more usual removable clothing. The Manhattan Doll Co. also marketed an unusual soldier walking doll. He was 28" tall and may have been made by the Babs Manufacturing Corp. in Philadelphia, Pennsylvania, circa 1917-1919. The doll also came dressed as a sailor. Other firms marketing soldier dolls during the era included the Elektra Toy and Novelty Co. located in New York City and the Horsman firm. For those wanting to sew handmade soldier, sailor, or nurse outfits for dolls, *The Delineator* offered patterns in a wide variety of sizes. In addition to strengthening their composition doll production, the E. I. Horsman firm also experimented with bringing American made bisque head dolls to the market. According to Don Jensen writing in *Collector's Guide to Horsman Dolls 1865-1950*, Edward Horsman Jr. contracted with Fulper Pottery of Flemington, New Jersey circa 1918 to work with Horsman to produce bisque doll heads. The dolls were introduced in June 1919 in four sizes. *Toys and Novelties* carried an ad for the dolls in September 1919. After a dispute with Horsman, the Fulper firm continued to produce heads on its own for several additional years. These dolls were sold to a variety of doll firms. The company manufactured socket heads for both baby and little girl dolls that were used with bent leg baby bodies or ball-jointed bodies, shoulder heads to be placed on kid bodies, and to a lesser degree toddlers for straight leg bodies. The dolls came in a variety of sizes. Production of the heads ceased circa 1921.

Companies in Japan were also called on to produce bisque head dolls that could fill the void left by those once supplied by Germany. Morimura Brothers as well as others began to import bisque dolls from Japan. In their advertisement in *Toys and Novelties* for September 1919, the Morimura Brothers announced that production was limited and recommended that businesses place orders early. Even years after the war ended, when Germany had re-established their doll exports, Japan continued to produce a variety of bisque dolls, usually of an inferior quality. Many of their war-time bisque dolls were based on molds from the German Armand Marseille firm, as were some of the Fulper dolls.

In addition to copying the bisque head dolls from Germany, American doll companies also produced doll heads made of composition that could be used on ball-jointed bodies. These dolls were made to look as much like their German counterparts as possible, even without the unavailable bisque heads. Many different firms produced these dolls, including Louis Amberg & Son (Victory Doll), Art Craft Toy Products Co., Beck Manufacturing Co., Bester Doll Manufacturing Co., Century Doll Co., Colonial Toy Mfg. Co., Inc., Effanbee (Fleischaker and Baum), National Joint-Limb Doll Co., and Trego Doll Mfg. Co., Inc. These dolls themselves make an interesting doll collection and provide reminders of a period in American history now nearly forgotten.

Also produced were baby dolls, which used composition heads similar to the German bisque models. These were mounted on composition five-piece bent limb baby bodies also based on the earlier German designs.

Because the old American doll corporations had expanded and new firms had been founded to meet the demand for dolls when German products were no longer available, these companies were worried about their futures when the war ended. Both toy and doll companies united in 1919 to fund a $65,000 advertising campaign called "American Toys For American Children." Doll firms involved in the campaign included Louis Amberg & Son, Averill Mfg. Co., Colonial Toy Mfg. Co., Federal Doll Mfg. Co., Fleischaker & Baum, E.I. Horsman and Aetna Co., Jessie McCutcheon Raleigh, and the A. Schoenhut Co.

A full page ad in *Toys and Novelties* for September 1919 re-enforced this message with a drawing of Uncle Sam holding two children and their toys, along with the slogan "Buy Toys Made in U.S.A." Arcade Manufacturing Co. paid for the advertisement.

Although it took a few years, by 1921 the Sears catalog was once again filled with bisque head German dolls in a variety of styles and sizes, including little girls, boys, and babies. The "Mary Jane" Effanbee composition ball-jointed doll was still being sold, priced from $4.95 to $8.95 for sizes from 16" to 24", while an imported German doll with a bisque head sold for only $5.95 in a 23" size. Many of the new doll firms could not survive the competition and disappeared from the scene shortly after the war ended. Others, like Amberg, Averill Mfg. Co., Effanbee, Horsman, Ideal, and Schoenhut, continued to produce dolls and prospered. The coming of the "Ma-Ma" doll (introduced by the Averill Mfg. Co.) and its popularity headed the doll market by 1923 and the American firms that had survived the years after the end of World War I met with new success. Not for many decades to come did American doll manufacturers have to endure heavy competition from overseas companies. There was still fierce competition in the doll industry, but now the U.S. firms were matched against each other. Only the giants would make it through the Depression and World War II.

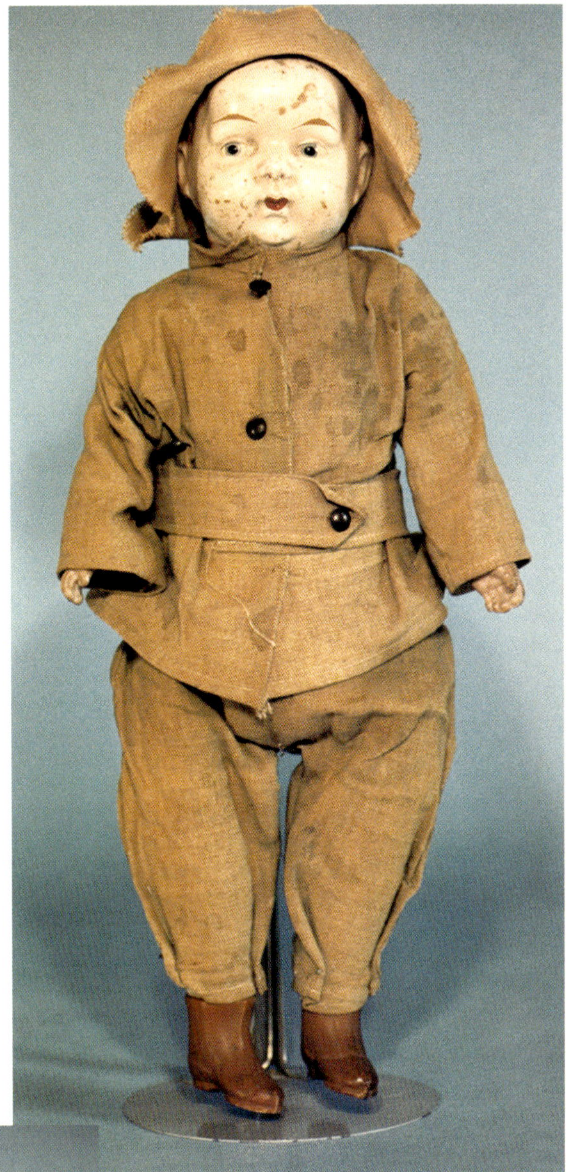

14" tall U.S. World War I soldier doll wearing his original uniform, circa 1916. The doll has a composition flange neck head, composition arms, cloth body, and made on boots. He has molded hair and painted features. His two-piece uniform retains its original buttons. Although the doll is unmarked, the same molded boots were used as footwear for many dolls made by Ideal. ($225+). *Silverthorn Collection.*

17" tall U.S. World War I unmarked soldier (missing hat), circa 1915-1916. The doll has a composition head with a flange neck, gauntlet lower composition arms, and a cloth body, upper arms, and legs. The soldier has painted molded hair and eyes and a closed mouth. He is wearing his original two-piece uniform but missing his original hat. ($175+). *Photograph and doll from the collection of Suzanne Silverthorn.*

12" tall, all composition World War I soldier boy made by the Ideal Toy & Novelty Co., circa 1918. He is sometimes called "Liberty Boy." The doll has painted features, molded hair, and is jointed at the shoulders and hips with steel springs. His uniform is molded on his body. The doll is missing his felt hat. He is marked on the back "U S/IDEAL" (in a diamond) of A." ($250). *Silverthorn Collection.*

Advertisement for the Manhattan Doll Co., Inc., whose showroom was on 5th Avenue in New York City. The ad appeared in *Toys and Novelty* magazine for September 1919. It features a 28" tall "Walking Doll" made of wood and composition, which walked when led by the hand. The retail price for the doll was $12.00 when dressed as pictured with a long curly wig. The doll also came with molded hair dressed as a soldier, sailor, or boy, priced at $10.00. The Colemans, writing in *The Collector's Encyclopedia of Dolls*, state that the Manhattan Doll Co. was in business as importers and jobbers from 1890 until 1922. *Ad from the collection of Marge Meisinger.*

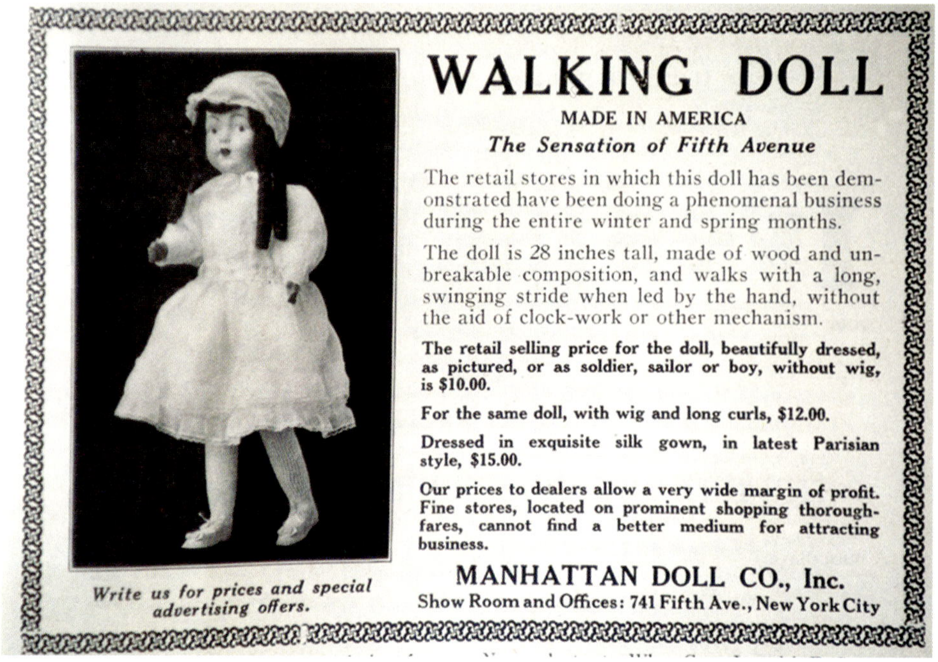

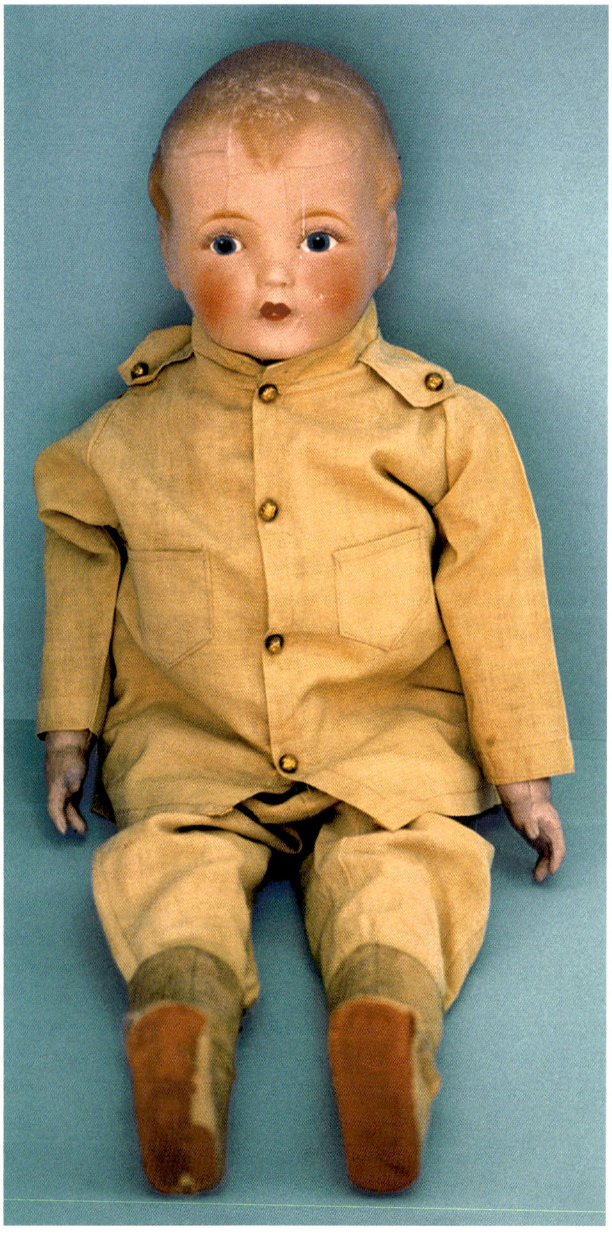

28" tall World War I soldier walking doll, circa 1918-1919. He has a composition shoulder head with molded hair and painted features. His hands are also composition. The top of the body is cloth while the rest of his body is made of wood pieces almost like those used on a marionette. The lower arms and legs are wood. The elbows and knees are hinged for walking. The doll appears to have once had springs at his knees as well as hinges. He wears his original World War I uniform, although he is missing his hat. Could he be the walking soldier doll advertised by the Manhattan Doll Co.? He is a walker and his size is right. His body is entirely different than the ones used on the Dolly Walker dolls. He may have been produced by the Babs Manufacturing Co. in Philadelphia, Pennsylvania, who manufactured 28" tall walking dolls from 1917-1920 with the Manhattan Doll Co. doing the marketing. Research will need to continue to solve the mystery. ($300+). *Doll from the collection of Ellyn McCorkell.*

21" tall Nurse doll made by the American Character Doll Co. located in New York City. Nurse dolls were very popular, both during World War I and shortly after. This doll probably dates from the mid 1920s. She has a composition shoulder head with full composition arms and legs and a cloth body. The doll has sleep eyes, open mouth with teeth, and a real hair wig. She is marked "Petite" on the back of her shoulder head. Her clothing and shoes appear to be original. She wears a striped cotton dress along with a nurse's apron, hat, and Red Cross pin. ($250+). *Courtesy of McCorkell Collection.*

18" tall bisque shoulder head doll made by the Fulper Pottery Co. in Flemington, New Jersey, circa 1919-1921. She is on a pin jointed leather body with lower composition arms and legs. The back of her head is marked with the vertical FULPER mark enclosed in an elongated circle. She has set blue eyes, open mouth with two teeth, and a real hair wig. She has been redressed in a dress made with old materials. ($200+).

Advertisement by the E.I. Horsman and Aetna Doll Co. for bisque dolls, which appeared in *Toys and Novelties* trade magazine in September 1919. The bisque heads were made by the Fulper Pottery Co., located in Flemington, New Jersey, to replace the popular imported German heads that were no longer available because of World War I. *Ad from the collection of Marge Meisinger.*

24" tall bisque socket head baby doll produced by the Fulper Pottery Co. The doll has a composition bent leg baby body. Many of the Fulper baby dolls have heads that appear to be too large for the body, as this one does. She has brown set eyes, open mouth with teeth, and a replaced mohair wig. The doll is marked with the vertical Fulper mark on the back of her head, plus "Made in/USA/16." She has been redressed in vintage baby clothing. More unusual Fulper bisque dolls include those with molded hair, including a Kewpie and a toddler. ($300+).

The Japanese import house called Morimura Brothers carried this advertisement in the September 1919 *Toys and Novelties* magazine. Several Japanese firms, including Morimura Brothers, marketed both little girl dolls and baby dolls during this period. White bisque dolls and celluloid dolls "Made in Japan" were also popular in the United States for many years. *Ad from the collection of Marge Meisinger.*

TOYS AND NOVELTIES

MORIMURA BROS.

53 West 23rd Street, New York

Announce Their 1920 Import Line

DOLLS—**Bisque Head**—See announcement on opposite page.
Bisque Babies—Complete line in staple and fancy designs.
China Limb Dolls—Repeating our excellent line of 1919.
Celluloid Dolls—Straight limb and celluloid babies in all sizes. Our "Dolly Doll" in the bathing suit. New novelty doll in fancy costume.

TOY TEA SETS—"The Little Hostess Tea Set" in display boxes, with new designs and packing. They are instructive as well and have found popular approval by the trade. We are sole distributors for this line.

White body china with hand painted designs—In white and gold, "Bluebird"—solid colors and children character designs. The most beautiful decorations ever shown. The quantity available of this line is very limited, therefore we advise early purchase.

XMAS TREE ORNAMENTS—A wonderful line with many new items and shapes.

CELLULOID NOVELTIES—A large line consisting of trumpets, floating animals, roly polies, funny figures, etc.

PAIER MACHE ANIMALS—Horses, dogs, cows and donkeys—On wood platforms and wheels. Natural colors and trappings. Lifelike. All sizes. Also sitting dogs, cats and mon'keys with voice.

SCHOOL BOXES—A new line of hardwood slide boxes.

NOVELTIES—Toy nursing and sewing sets, fur dogs and a varied line of new items.

LACQUER AND BASKETWARE

Production is very limited—Materials and labor very uncertain. We therefore suggest that you anticipate your 1920 requirements and place your orders early with our representatives who will call on you shortly, or by seeing our sample display in New York.

WHEN ANSWERING ADVERTISEMENTS, PLEASE MENTION TOYS AND NOVELTIES

12" tall Japanese Morimura Brothers baby doll with bisque socket head on a five-piece bent leg baby body, circa 1918-1919. The head has set glass eyes, open mouth with four teeth, and a replacement wig. She is marked on the back of the head in a circle "MB" with a simple design between the letters. "Japan" and the number "2" are also included in the marking. She has painted eyelashes all around the eyes and a dot in the inner corner of each eye with matching red dots marking her nostrils. She has been redressed. ($150+).

24" tall Japanese Morimura Brothers bisque socket head doll with ball-jointed body, circa 1917-1918. She has painted eyelashes on the bottom and real eyelashes at the top of her eyes, brown blown glass eyes, open mouth with teeth, and a replaced long curl wig. She is marked on the back of her head with "MB" in a circle and "Japan/7" under the MB marking. The doll has been redressed. The doll chair she is sitting in is circa 1910. (See "Furniture" chapter for more of this furniture.) ($200-225).

Advertisement for Amberg Dolls featured in *Toys and Novelties* trade magazine from September 1919. The ad from the Louis Amberg & Son New York firm lists the firm's current dolls, which include the Victory Doll featuring a composition socket head on a ball-jointed body. Another doll from the ad was Amberg's Walking Doll. *Meisinger Collection.*

A close-up of the Amberg Victory doll, picturing her tag and chemise. She still has her two original pink hair ribbons. The other one is in the back of her hair. Her original real hair wig is still in its original set that features long curls.

15.5" tall Victory doll, made by the Louis Amberg & Son firm of New York City. She has a composition socket head, ball-jointed body, sleep eyes, open mouth with teeth, and a real hair wig. She is dressed in her original chemise and has her tag and box. She is marked on the back of the head "L.A. & S/40." Her back is marked "Amberg's/Victory/Doll." Her tag reads "Amberg's /Victory/ Doll/Made in U.S.A." The other tag information deals with care of her eyes. Her box information reads "Amberg Doll/The American Standard/ Plastic Model Copyrighted Louis Amberg & Son 1912/Trademark Registered/Truest-To-Life/Hardest-To-Break." The model number looks like a "P. 1500" written in pencil. ($200+).

Artcraft Toy Products Company advertisement that appeared in *Toys and Novelties* magazine in September 1919. One of the firm's ball-jointed dolls was pictured. The copy read "The Artcraft doll combines the most desirable qualities of the German doll with the wonders added by the superior creative skill of American artists." The dolls had wood-pulp composition bodies except in the larger models, which had papier mâché bodies. The dolls came in 20", 23", and 26" sizes. The firm also produced baby dolls. *Meisinger Collection.*

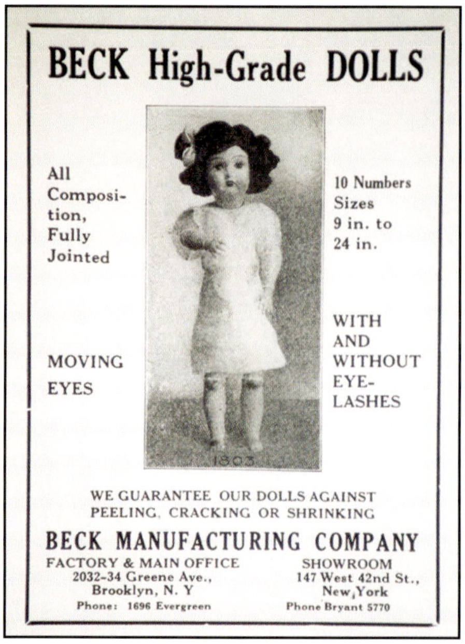

Advertisement for the Beck Manufacturing Co. of Brooklyn N.Y. from *Toys and Novelties* trade magazine of September 1919. The dolls were all composition and included fully ball-jointed bodies and sleep eyes. They ranged from 9" to 24" tall. *Meisinger Collection.*

20" tall composition socket head doll made by the Artcraft Toy Products Co., circa 1919-1920. The doll has sleep eyes, open mouth and teeth, and a ball-jointed body. Her wig has been replaced. She is marked on the back of her head in a half circle "ART/CRAFT" and under that "U.S.A." The doll has been redressed. ($100 with some wear).

16" tall composition socket head doll made by the Bester Doll Manufacturing Co. of Bloomfield and Newark, New Jersey, circa 1918-1919. The doll has a ball-jointed body, sleeping eyes, open mouth with teeth, and her original real hair wig. She is marked "Bester Doll/Co." and other words that cannot be identified. The dolls were made in sizes from 16" to 26" tall. This doll has been redressed. ($200).

19" tall doll with ball-jointed body, made by the Century Doll Co. in New York City. The back of the composition head is marked "Century/ Doll Co." She has blown glass blue eyes, closed mouth, and a real hair wig (may be a replacement). She has been redressed. (With paint flecking, missing thumb $75-100).

TOYS AND NOVELTIES

Colonial Dolls

MADE IN AMERICA BY AMERICANS

WHEREVER good dolls and toys are sold there you will find COLONIAL DOLLS. Each Colonial Doll is built to a jealously guarded standard of excellence by an organization trained to know only the best.

Colonial Toy Mfg. Co., Inc.

Advertisement for the Colonial Toy Mfg. Co., Inc. in the *Toys and Novelties* magazine for September 1919. The New York based firm was in business during and after World War I. The ad pictures one of its composition head dolls on a ball-jointed body and a socket head baby doll on a five-piece baby body. The doll dressed in the Colonial costume was also used as a trademark for the firm. Baby dolls were manufactured with both painted and sleeping eyes and with molded hair or wigs. In 1916, "Zaiden" was part of the mark on the back of the heads. *Meisinger Collection.*

14" tall composition doll made by the Colonial Toy Mfg. Co., circa 1916. He has a five-piece jointed curved-limb baby body, molded hair, painted eyes, and a closed mouth. The doll is marked on the back of the head: "Colonial/Zaiden/Doll." On the back of the body is: "Zaiden/ Doll/Colonial Toy MFG Co." ($100).

Effanbee Mary Jane doll advertisement that appeared in the Sears, Roebuck catalog, circa 1920. She was an all composition doll that included a ball-jointed body like those used on the German dolls with bisque heads. She came in sizes of 16", 18", and 24" priced at $6.25, $7.15, $8.33, and $10.48. The completely dressed doll had moving eyes and a real hair wig. She was jointed at the neck, shoulders, elbows, wrists, knees, and hips.

16" Mary Jane doll made by Effanbee, circa 1919-1921. The doll has a composition head, a ball-jointed body, sleep eyes, a real hair wig, and an open mouth with teeth. She is marked "Effanbee" on the back of her head and the back of her body. She wears a vintage handmade dress and underwear, probably made for her when she was new. Her shoes and socks have been replaced. Effanbee also produced a doll at this time with a composition shoulder head on a pin-jointed leather body. Her arms were ball-jointed composition, as were her lower legs. She had sleep eyes and an open mouth with teeth. (Damaged foot $135).

18.5" tall doll with ball-jointed body made by the National Joint-Limb Doll Co. of New York City, circa 1917-1918. She has a composition socket head, sleep eyes, open mouth and teeth, with a papier mâché body. Her mohair wig may be original. Both the back of her head and the back of the body are marked "NAJO" in a double circle. The back of the head is also marked "19C." The dolls came in sizes from 18" to 26" tall in 1918. The firm also produced baby dolls with bent legs and with straight legs. The doll has been redressed. ($135-150 with some wear).

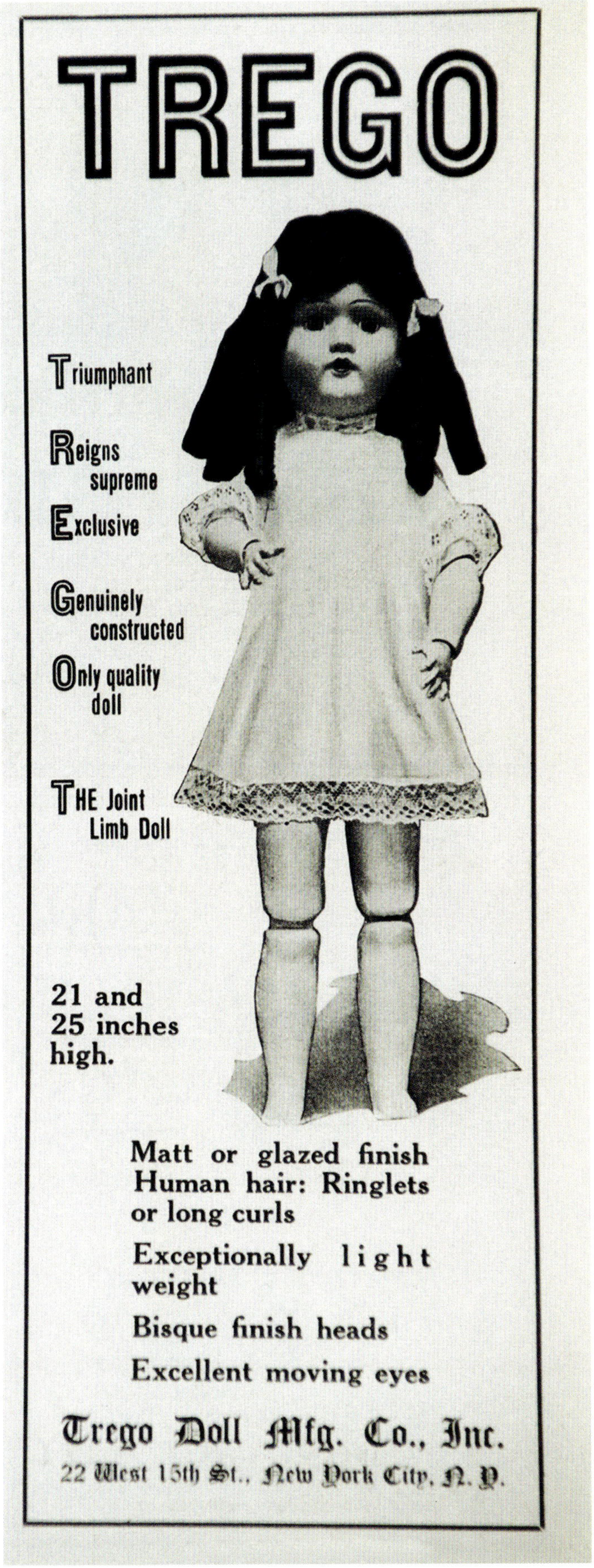

Advertisement for the Trego Doll Mfg. Co. in *Toys and Novelties* magazine for September 1919. The New York based firm offered composition socket head dolls on ball-jointed bodies with sleeping eyes in sizes of 21" and 25". The dolls came with human hair wigs and were dressed only in a chemise. *Meisinger Collection.*

21" tall Trego composition socket head doll on ball-jointed body. She has sleep eyes, open mouth with teeth, and her original real hair wig. She is marked on the back of her body "TREGO/ Made in U.S.A." She has been redressed. ($175+).

This advertisement was featured in the September 1919 *Toys and Novelties* magazine. Many American toy manufactures were afraid that their customers would return to former German toy suppliers since World War I was over, and they banded together to launch a publicity campaign designed to keep this from happening. Their slogan was "Buy Toys Made in U.S.A." Although the Arcade Manufacturing Co. was responsible for this ad, many doll companies contributed money for the campaign. The same magazine featured an article titled "American Toys for American Children," detailing the $65,000 campaign to "Buy American." Doll companies on board included: Louis Amberg & Son, Averill Mfg. Co., Colonial Toy Mfg. Co., Federal Doll Mfg. Co., Fleischaker & Baum (Effanbee), E.I. Horsman & Aetna Co., Jessie McCutcheon Raleigh, and A. Schoenhut Co. *Meisinger Collection.*

Chapter 10

Furniture and Related Accessories

The period from 1910 through 1930 offers many different designs of doll furniture and other accessories for today's collectors. The most popular pieces include doll beds and cradles, carriages and carts, dressers and trunks, rockers and chairs, stoves and cabinets, and tables and chairs.

From 1910 until 1920, most of the doll furniture was ornate, with added decorations of balls, beaded spindles, or other ornamentation. The furniture usually came with either enameled, oak, or maple finishes. Kitchen pieces reflected the styles of the day with heavy iron cooking stoves and oak kitchen cabinets. Doll clothes were stored in cedar chests or trunks, although the camelback trunks had lost popularity by this time. Dolls were transported in reed carriages (although the elaborate models that included umbrellas were no longer in fashion), folding carts made of steel and imitation leather, or perambulators with wood bodies.

By the 1920s, doll furniture had become much more streamlined. The pieces were designed with boxy straight lines with little more than a decal or two used for decoration. Most of the furniture came with enamel finishes, often in various shades of green. Although cradles were still available for purchase, the new doll cribs on wheels seemed to be more popular. Mission oak rockers were also available during this period, in a variety of child sizes. These rockers can be used to provide attractive seating for vintage collectible dolls. Kitchen pieces in the 1920s changed considerably from those of the earlier decade. The stoves were made with long legs to reflect the 1920s look. Some of these stoves were made to cook with electricity. The cabinets also reflected a more modern look. Doll clothing was stored in more modern trunks (with flat tops) throughout the decade. Carriages continued to be made of round fiber for the most part, but doll buggies made with metal bodies and folding hoods were also offered. Doll strollers and sulkies could be purchased at cheaper prices to provide additional transportation for dolly. One model made of round fiber sold for $1.98. The metal sulky was priced at only 89 cents.

Doll dishes have changed less than doll furniture over the years. Although the china dishes from Germany were originally higher priced than the metal tea sets from Ohio Art, both types of dishes from the period remain very collectible. German dishes became scarce during the World War I years, so Japan began exporting china dishes to the United States to help fill the void. They continued this practice for decades, except for the time during World War II. Although most of these exports were not high quality products, they did fill a void by providing inexpensive china tea sets for many years.

There are so many wonderful vintage beds, buggies, dressers, chairs, trunks, and kitchen pieces still available from 1910-1930 that the only problem for collectors is finding the space to display them adequately. Space for a large doll sitting in a child's rocker from the period can usually be found in a living room, family room, or bedroom. The dolls look right at home in any of these rooms. For the collector who is lucky enough to have a doll room, floor or shelf space should be available for several doll beds, cradles, carriages, dressers, or chairs. For others, a shelf or shelves attached to a wall can be used to make an arrangement using a small doll dresser, open trunk with clothing, and a small folding bed from the period. The area sometimes left above kitchen cabinets also allows space to display a variety of vintage kitchen pieces, including stoves, sinks, and washing machines. In addition, some modern houses include wide ledges near the ceiling that provide space to display larger doll related items such as doll carriages and bigger doll kitchen cabinets. Collectors will need to choose from these accessories carefully and then be creative in their display plans.

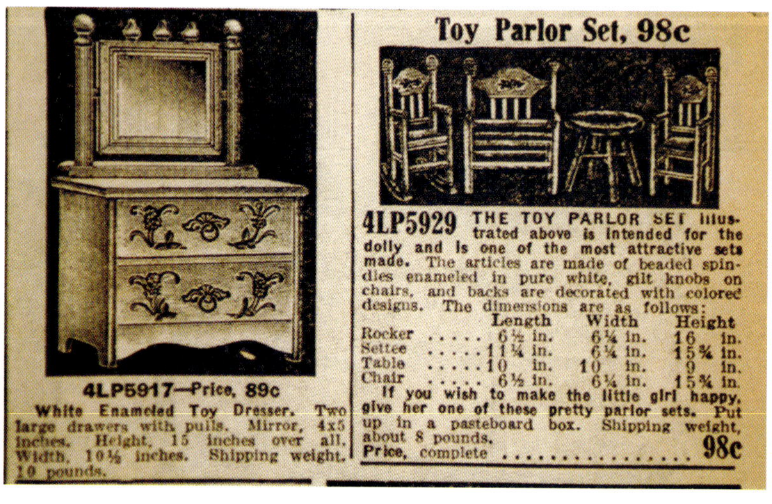

Advertisement for a doll dresser and toy parlor set carried in the Montgomery Ward catalog in 1914. The design of this furniture was probably based on the stick and ball adult furniture from the period. The doll dresser was finished in white enamel and included two drawers. It measures 15" high x 10.5" wide and sold for 89 cents. The parlor set included a settee, rocker, chair, and round table. The entire set sold for only 98 cents. A matching bedroom set was featured in the Sears Roebuck catalog for 1912. It included a bed (18" long), rocker, chair, (13.5" tall), round table, and clothes rack 24" tall. The set sold for $1.45. *Catalog from the collection of Marge Meisinger.*

Toy parlor set exactly like the one in the 1914 Montgomery Ward catalog but missing the table. The chair and settee are 15.5" high while the rocker measures 16" tall. The settee is 11.25" wide. Dolls 15" to 16" tall can be used with the furniture. This set includes its original paint and decal decoration. ($125-150 set).

Children's Dining Set

4LP5907 Child's Dining Set. Consists of one table 15½ inches high with top, size 10x20 in.; two chairs, 22 in. high; seat, 10¼x10 in. All made of seasoned wood and finished in pure white enamel with gilt knobs and colored decorations on chairs. These sets are large enough and strong enough for practical use by the children. Each set packed in strong crate. Shipping weight, 20 pounds.

Price, set complete $1.75

The same Montgomery Ward 1914 catalog also advertised a matching table and two chairs in a larger size. The chairs were 22" high and the table was 15.5" tall. The set sold for $1.75. *Meisinger Collection.*

This doll dresser is similar to the one pictured in the Montgomery Ward catalog, but it is a larger size. This dresser has three drawers and is 25.5" high, including the decorations on top of the mirror, and 16" wide. It dates from the same time period. ($100+).

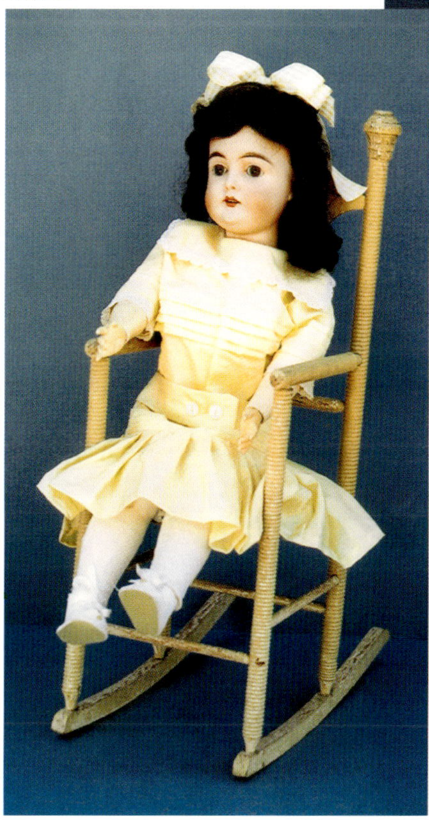

A larger rocking chair was produced by the same company to match the chairs in the previous 1914 ad. Pictured is one 23.5" tall (the rockers make it a little higher than the straight chairs). This rocker has been repainted. This chair will accommodate a ball-jointed doll approximately 23" tall. (Repainted $60).

The same firm also made this high chair. All of this type furniture came with the knobs painted gold. The high chair measures 28.5" tall and 8.5" wide, not including the tray measurement. All of this line of furniture is designed with a tall, thin look. The high chair trays are supposed to lift over the doll's head but there is not enough space to use this feature effectively for larger dolls. Perhaps that is why many chairs are found with missing trays. They broke or were removed to accommodate a larger doll. This tray has been repainted.($100).

Child's size rocker made with the stick and ball design, circa 1890-1910. Like the earlier pictured doll furniture, this chair includes knobs on the top and beaded spindles. It measures 25" high. ($125).

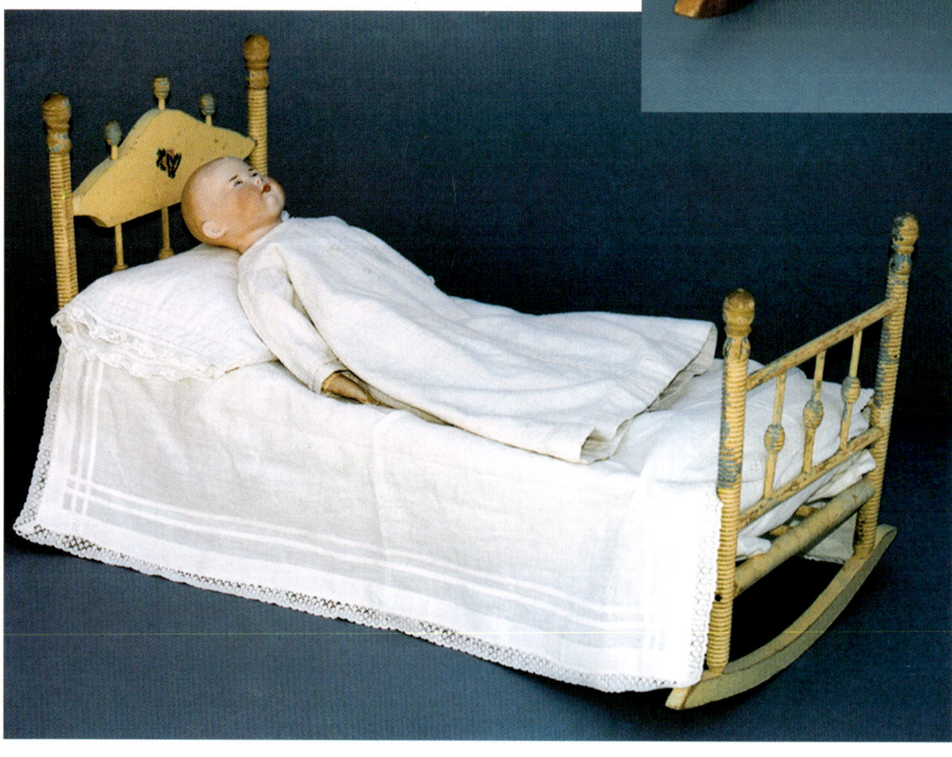

The cradle, which seems to match the earlier furniture, was sold for many years, both in this design and related styles. This example includes the decal decoration as well as the corner knobs. The cradle is finished in cream enamel with blue knobs. It measures 15.5" high x 12.5" wide x 26.5" long. ($75-100).

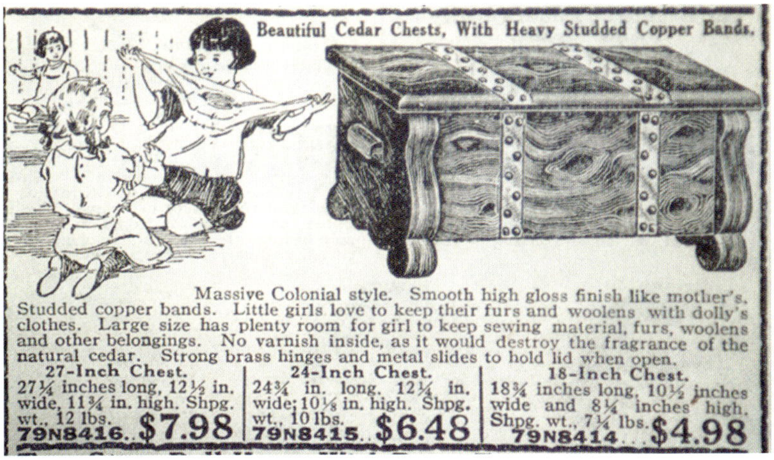

During the early years of the twentieth century, small cedar chests were produced especially for little girls. Most of these chests were used to hold doll clothing. This ad is from the Sears 1923 Fall & Winter catalog. It pictured a chest that came in 18", 24", and 27" sizes. The cedar chests were priced from \$4.98 to \$7.98 each.

Collapsible doll cart similar to the ones pictured in the Montgomery Ward catalog. The top, back, seat, and footrest were finished in imitation leather (top is missing its original covering). The metal frame is black. The cart is 21" high to the handle, and 15" long. (\$75).

Cedar chest, circa 1915-1920, owned by the author's mother during her childhood. The chest was also used for storage of doll clothes when the author was a child. It measures 11.5" high x 14.5" wide x 10.75" deep. (\$50+).

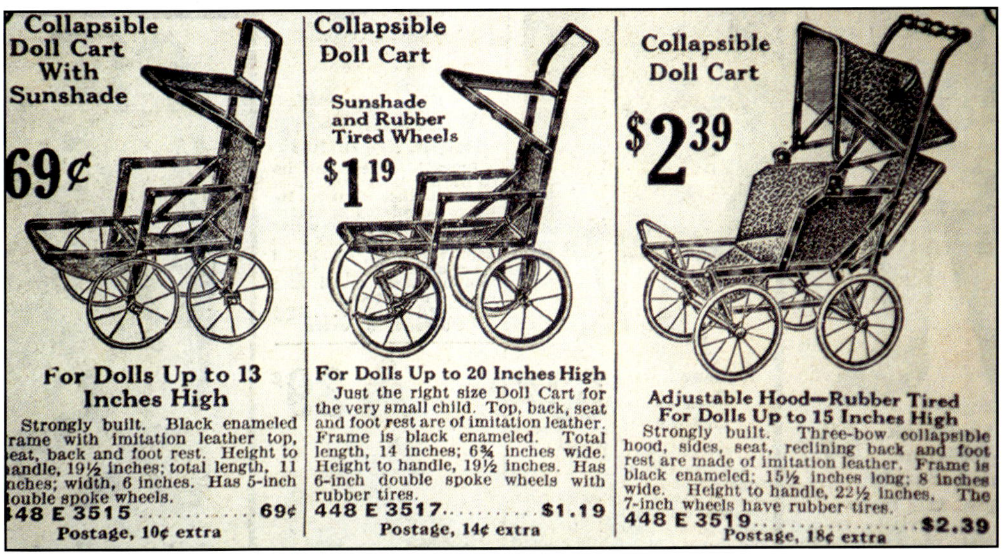

This Montgomery Ward catalog ad from 1924 pictures three different styles of doll carts. A similar folding doll "cab" was featured in the 1916 Montgomery Ward catalog priced at \$2.98. The 1924 price was \$2.39. The two cheaper carts were priced at 69 cents and \$1.19. These carts had been advertised since 1917 or before.

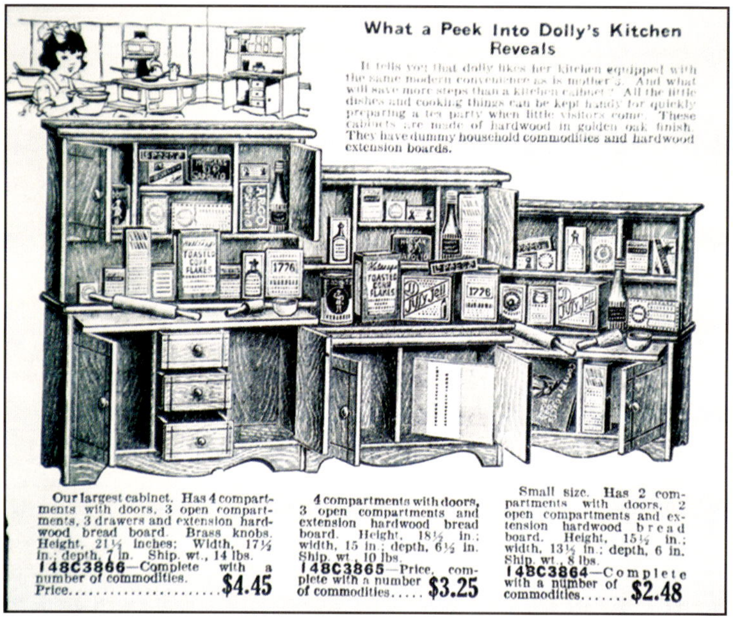

Advertisement for oak kitchen cabinets made by N.D. Cass shown in the Montgomery Ward catalog for 1920. The cabinets came in at least four sizes; the smallest is not pictured. It was 13" high x 11" wide and sold for $1.19 including groceries in 1914. The largest model was 21.5" tall x 17.5" wide and was priced at $4.45 in 1920. It also included three drawers. The cabinets had been advertised as early as 1914 by Butler Brothers. *Courtesy of Ronald S. Barlow from his book The Great American Antique Toy Bazaar 1879-1945.*

Oak kitchen cabinet made by the N.D. Cass firm, circa early 1920s. It is 13" high x 11.5" wide. The cabinet includes two opening doors, a pull out extension bread board, and an extra shelf. It still retains its original stamped marking that reads "Cass/Kitchen Cabinet." It originally came with cardboard grocery boxes, a rolling pin, and other utensils. The Britannia metal teapot, cream, and sugar with a silver finish are circa 1917. (Cabinet $100).

The Montgomery Ward catalog for 1920 also advertised a variety of cast iron toy cook stoves. They ranged in price from 79 cents to $6.25 each. The largest model measured 13.5" tall x 15.5" wide and sold for $6.25. The smallest was 5.5" tall x 4.75" wide. It was priced at 79 cents. This small iron cook stove labeled Eagle was produced by the Hubley Manufacturing Co., located in Lancaster, Pennsylvania circa 1920s. It measures 3.25" tall (not including the stove pipe), 5" wide, and 4" deep. The oven and lids open. ($100+). *Photograph and stove from the collection of Gail and Ray Carey.*

Right:
"Bissell's Midget" toy carpet sweeper. It is marked "Cyco Ball-bearing Sweepers For the Home." The Bissell firm produced different designs of these sweepers for a number of years. Each design included an individual name. The 1914 Marshall Field Co. catalog featured "Little Daisy, Little Gem, Little Jewell, and Misses" sweepers. This one is made of oak, has wood wheels, wood pans (underneath), and a bristle brush. The sweeper (not including the handle) measures 2.5" high x 7.5" wide x 5" deep.($75-100 in excellent condition).

This unmarked china tea set, circa 1915-1920, originally included six cups and saucers along with a teapot, creamer, and sugar. It was probably made in Germany. The saucer measures 3" in diameter, the teapot is 4.5" tall. The dishes are decorated with pink flowers and gold accents. ($75+).

Three drawer dresser with mirror, featuring golden oak finish, circa 1912-1920. Other pieces of doll furniture were also produced in this style. Included were china cabinets, washstands, sideboards, chiffoniers, kitchen cabinets, and beds. The dresser is 15.5" high (to top of the mirror) x 9" wide x 5" deep. The original price marked on its back in pencil is 35 cents. It includes three drawers (not made of oak) decorated with pressed designs, as is the mirror frame. The knobs are wood. (Dresser $55+).

Below:
This washstand, wardrobe, and bed are circa early 1900s to 1915. The washstand and wardrobe match. Both have decorative pieces added that are finished in a red stain. The one on the washstand is the splash board while the wardrobe's red stained piece is across the bottom. Both items include wood knobs and decorative grooves in the wood. The wardrobe originally came with a bed that folded up into its back, so the drawers are only 2.75" deep. The washstand measures 6.5" high x 5.75" wide x 2.75" deep. It includes three drawers and is marked on the back in pencil 10 cents. The wardrobe measures 13.25" high x 10" wide x 4.25" deep. The folding metal bed is 14" tall x 14" long x 9" wide. The 1914 Marshall Field & Co. catalog features similar beds. Their description reads "Folding wire doll beds with brass finish, head, foot, and sides all folding. Made of round steel wire strongly jointed: fold flat." Their beds were trimmed with cretonne and came with a mattress, pillows, and head board trim. (Washstand $35, wardrobe $65, bed $50).

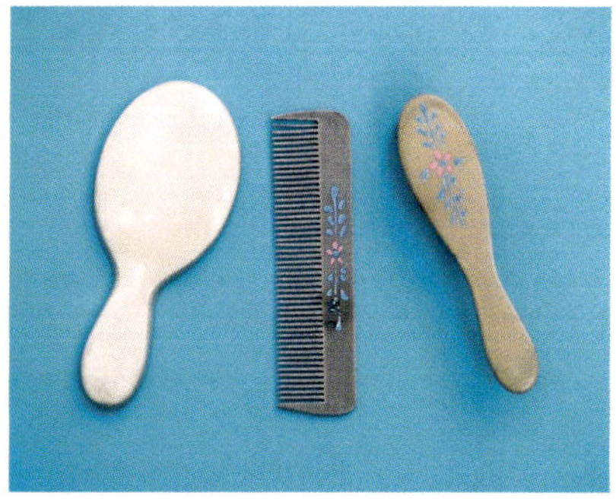

Celluloid doll comb, brush, and mirror similar to those pictured in the Sears catalog in 1922. Sears also included a larger set (soap, powder, and rouge boxes added). The mirrors in the Sears examples were 4" high while the pictured mirror is 3.25" tall. ($5-10 each).

Wood shoofly, circa 1915. It features two wood horses connected with a seat in the middle and measures 19.5" high x 32.5" wide x 15.5" deep. The horses swing back and forth. A bisque head doll with ball-jointed body enjoys the ride. (Not enough examples to determine a price). *From the collection of Jeff Zillner.*

0917/3. Assortment, Three Kinds. Arm Chair, height 10½ inches, seat 5x4¾ inches; Rocker, height 11¼ inches, seat 5x4¾ inches; High Chair, height 13¾ inches, seat 5x4¾ inches. All in white enamel finish; 1 dozen assorted pieces in pkg. ...... Doz. **$4.50**
In lots of 3 dozen ...... Doz. **$4.00**

Advertisement for three doll chairs from the Nerlich 1924 catalog. By the 1920s, doll furniture was being made with fewer frills and had become more square in design. This set included an arm chair 10.5" high, a rocker 11.25" tall, and a high chair 13.75" high (notice no tray).

A high chair similar to the one pictured in the Nerlich advertisement from 1924. This chair retains the knobs from the earlier time period so it may date from 1915 to 1920. It measures 15" high x 5.75" wide. Like the Nerlich example, it also has no tray. A Kestner baby featured in the Bisque Baby section occupies the chair. ($35).

Butler Brothers ad from 1929 featuring square white enameled doll furniture from the period. Included were a 14" tall two-drawer dresser with a mirror, a one-drawer dressing table with a three way mirror measuring 12" high, and a 16" tall china cabinet with two drawers and glass covered shelves. *Catalog from the collection of Marge Meisinger.*

Wood doll cribs on wheels advertised in the Fall-Winter M.W. Savage Minneapolis, Minnesota, catalog for 1929-30. Both cribs could hold dolls 25" tall. *Catalog from the collection of Marge Meisinger.*

17" tall dresser with two drawers and mirror made by N. D. Cass Co., circa 1920s. It is 12" wide and 5.75" deep and decorated with a decal design of a bird. Some of the other doll dressers of the period featured stencil decorations. It is marked "N.D. Cass Co./Made in US.A." under the bluebird label. Sears advertised a 20.5" high decal decorated dresser in their 1925 catalog priced for $2.59. ($75). *Photograph and dresser from the collection of Patty Cooper.*

Wood doll crib from the 1920s that includes wood wheels. The crib measures 20" high x 27.5" long x 15" deep. A similar crib 23.25" high was priced for $2.79 in the Sears catalog in 1923. ($50+).

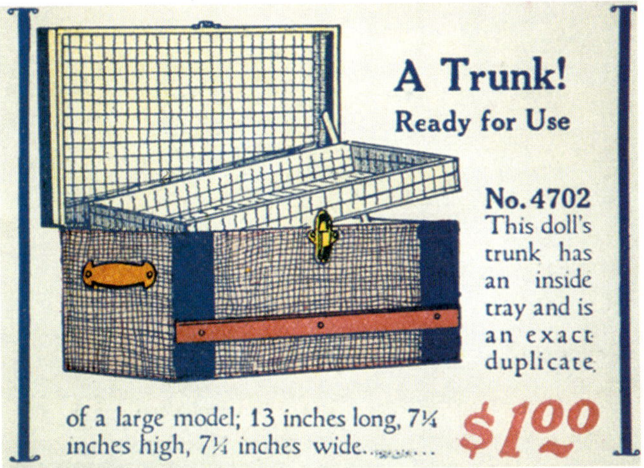

This doll trunk measuring 7.25" high x 13" long x 7.25" deep was advertised in the 1920s for only $1.00. It included a lift-out tray. Sears carried a similar trunk in their 1921 catalog priced for $1.39.

Doll trunk made of wood with paper covering, circa early 1920s. It includes a removable tray and a lock on the front. The trunks were used to store doll clothing. This one measures 5" high x 10" wide x 6" deep. Pictured with the trunk are two dolls from the period. The unidentified baby has a bisque head and papier mâché body. The little girl doll was made by the German Ernst Heubach firm. ($40).

Furniture for dolls and/or small children was also made of wicker during the 1920s and 1930s. Round fiber was often used in the production of these pieces. The pictured rocking chair dates from the 1920s. It is 23" tall, 12" wide, and nearly 13" deep. A large Dolly Walker doll rests comfortably in the chair. ($50-75).

A matching wicker table and chair also came in the set. The three pieces had remained with their original owner for many years and have always been together. All three pieces recently received a new coat of paint. The table and chair are especially unusual when made of wicker. The chair is nearly 22" tall, 11" wide, and 11.5" deep. The table top measures 18" square and is 18" high. A Dolly Walker doll gets ready to sip his tea at the table. The china tea set is unmarked (Set with repainting $100-125).

Advertisement for an "Extra Large Size" doll carriage from the 1924 Montgomery Ward catalog. The carriage was made of round fiber with corduroy upholstery. It had rubber tires with nickel plated hub cabs. The unusual feature of the buggy was its color – Royal Blue. It measured 29.5" high (to top of handle), 34.75" to top of the hood, 25.5" for the length of the body, and 12" wide. The carriage also had a footbrake. The same model was also pictured in tan in later catalogs.

Doll carriage that looks very much like the one pictured in the 1924 catalog in size and style. It is also made of round fiber and is upholstered with corduroy. The carriage has been repainted but still retains its footbrake. It measures 33" to the top of the hood, the length of the body is 26", and it is 13" wide. A Fulper baby doll is pictured in the carriage. ($100-125 repainted).

Metal doll sulky and cart made of wicker, both used for doll transportation. The sulky was sold by Sears beginning in 1922, when they were priced for 89 cents each. The handle is approximately 20.5" long with the seat measuring 7" x 7". The wicker cart handle measures 20" long and the seat is 9" wide x 7" deep. Its handle and two posts were made with "beaded spindles" like the early furniture featured in the Montgomery Ward 1914 catalog. (Sulky $75, Wicker cart with some repair $40).

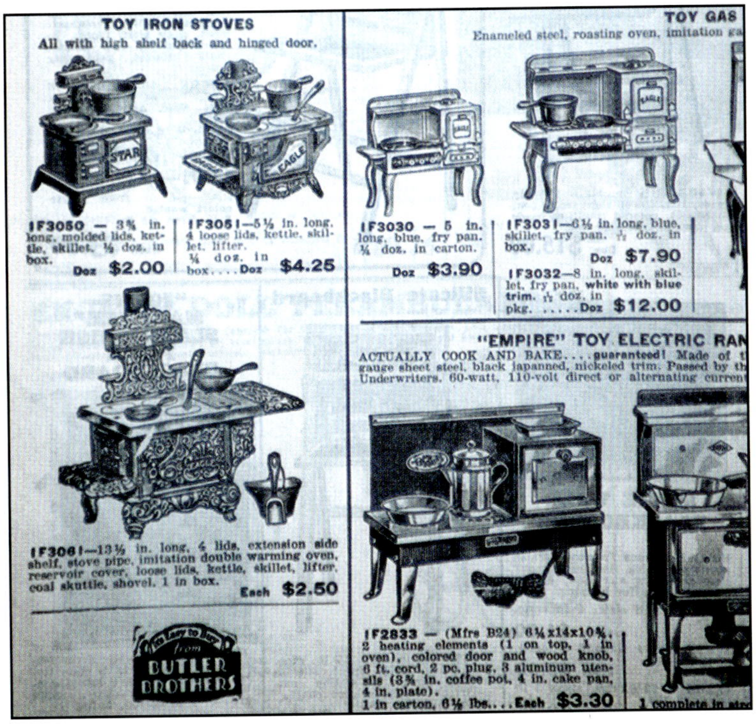

TOY IRON STOVES

All with high shelf back and hinged door.

1F3050 — 3⅜ in. long, molded lids, kettle, skillet. ½ doz. in box. Doz $2.00

1F3051—5½ in. long, 4 loose lids, kettle, skillet, lifter. ⅙ doz. in box....Doz $4.25

1F3061—13½ in. long, 4 lids, extension side shelf, stove pipe, imitation double warming oven, reservoir cover, loose lids, kettle, skillet, lifter, coal skuttle, shovel. 1 in box. Each $2.50

It's Easy to Buy from BUTLER BROTHERS

TOY GAS

Enameled steel, roasting oven, imitation ga

1F3030 — 5 in. long, blue, fry pan. ¼ doz. in carton. Doz $3.90

1F3031—6½ in. long, blue, skillet, fry pan. 1/12 doz. in box. Doz $7.90

1F3032—8 in. long, skillet, fry pan, white with blue trim. 1/12 doz. in pkg. .....Doz $12.00

"EMPIRE" TOY ELECTRIC RAN

ACTUALLY COOK AND BAKE....guaranteed! Made of t gauge sheet steel, black japanned, nickeled trim. Passed by th Underwriters. 60-watt, 110-volt direct or alternating curren

1F2833 — (Mfrs B24) 6¼x14x10¾. 2 heating elements (1 on top, 1 in oven), colored door and wood knob, 6 ft. cord, 2 pc. plug, 3 aluminum utensils (3⅜ in. coffee pot, 4 in. cake pan, 4 in. plate). 1 in carton, 6½ lbs....Each $3.30

1 complete in st

Advertisement for toy stoves from the Butler Brothers 1929 holiday catalog. Three types of stoves were featured, including several sizes of iron stoves, Eagle gas ranges, and the Empire electric stoves. *Catalog from the collection of Marge Meisinger.*

Metal ice box produced by the V. M. Kearley Manufacturing Co. of New York City, circa 1931. The toy is marked on the outside of the door with a decal reading "Another Kearley Creation" along with the company name. The "coil" on top is made of wood and the rest of the toy is made of metal. The inside includes two shelves and a place for the ice cube trays. The ice box measures 13" high x 7.5" wide x 5" deep. ($75+)

Cast iron kitchen furniture produced by the Hubley Manufacturing Co., circa 1920s. The ice box measures 6" high x 4.5" wide. The cabinet is 9" high and the stove is 6.5" tall. The doors are functional on all three pieces. The Hubley Manufacturing Co., located in Lancaster, Pennsylvania, produced stoves marked "Eagle," ice boxes labeled "Alaska," kitchen cabinets marked "Cabinet," and kitchen tables and chairs. The stoves came in sizes from 4.5" to 12.5" tall ($125 each).

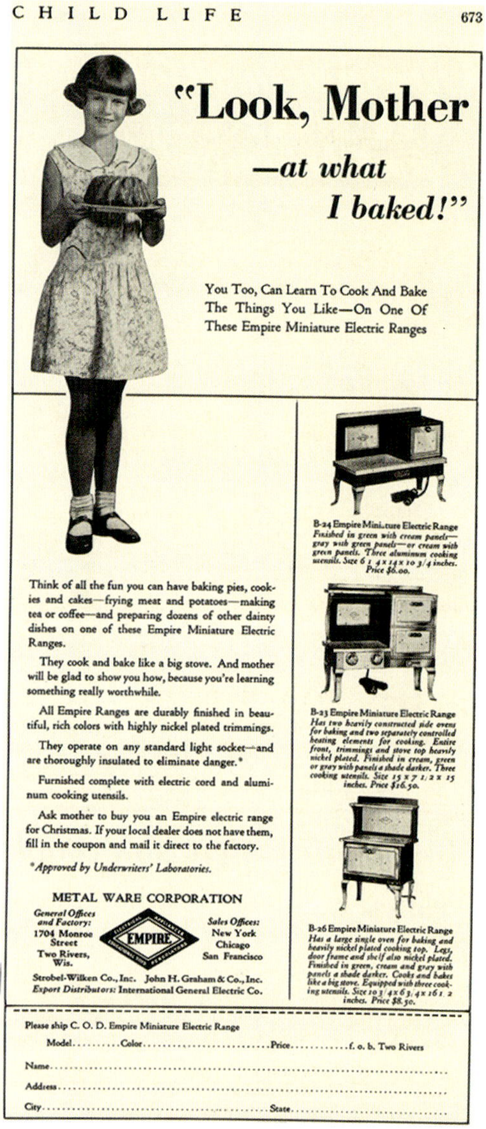

CHILD LIFE 673

"Look, Mother —at what I baked!"

You Too, Can Learn To Cook And Bake The Things You Like—On One Of These Empire Miniature Electric Ranges

Think of all the fun you can have baking pies, cookies and cakes—frying meat and potatoes—making tea or coffee—and preparing dozens of other dainty dishes on one of these Empire Miniature Electric Ranges.

They cook and bake like a big stove. And mother will be glad to show you how, because you're learning something really worthwhile.

All Empire Ranges are durably finished in beautiful, rich colors with highly nickel plated trimmings.

They operate on any standard light socket—and are thoroughly insulated to eliminate danger.*

Furnished complete with electric cord and aluminum cooking utensils.

Ask mother to buy you an Empire electric range for Christmas. If your local dealer does not have them, fill in the coupon and mail it direct to the factory.

**Approved by Underwriters' Laboratories.*

METAL WARE CORPORATION

General Offices and Factory: 1704 Monroe Street Two Rivers, Wis.

EMPIRE

Sales Offices: New York Chicago San Francisco

Strobel-Wilken Co., Inc. John H. Graham & Co., Inc.
Export Distributors: International General Electric Co.

B-24 Empire Miniature Electric Range *Finished in green with cream panels—gray with green panels—or cream with green panels. Three aluminum cooking utensils. Size 6 1/4 x 14 x 10 3/4 inches. Price $6.00.*

B-23 Empire Miniature Electric Range *Has two heavily constructed side ovens for baking and two separately controlled heating elements for cooking. Entire front, trimmings and stove top heavily nickel plated. Finished in cream, green or gray with panels a shade darker. Three cooking utensils. Size 15 x 7 1/2 x 15 inches. Price $16.50.*

B-26 Empire Miniature Electric Range *Has a large single oven for baking and heavily nickel plated cooking top. Legs, door frame and shelf also nickel plated. Finished in green, cream and gray with panels a shade darker. Cooks and bakes like a big stove. Equipped with three cooking utensils. Size 10 3/4 x 6 3/4 x 16 1/2 inches. Price $8.50.*

Please ship C. O. D. Empire Miniature Electric Range
Model.........Color..........................Price...........f. o. b. Two Rivers
Name..
Address...
City..................................State....................................

Empire Stove advertisement that appeared in the December 1930 issue of *Child Life* magazine. It pictures three different models of toy electric stoves made by the Metal Ware Corp. of Two Rivers, Wisconsin. The stoves ranged in price from $6.00 to $16.50. The most expensive model included two working ovens.

Empire toy electric range called model B-26 in the Metal Ware Corp. ad. It has a large single oven for baking and a two burner nickel plated cooking top. It originally sold for $8.50. The metal label on the bottom reads "Empire Metal Ware Corp./Two Rivers, Wis." Another "Empire" label is located on the back board. The stove measures 16.5" high x 10.75" wide x 5.75" deep. The vintage tea-kettle is not original to the stove although the company advertising stated it came with three cooking utensils. ($125+).

Advertisement from the South Bend Toy Mfg. Co. in South Bend, Indiana that appeared in *Child Life* magazine in December 1931. The firm produced toy cabinets and child's tables and chairs as well as other products. The pictured cabinets included two glass doors on the top. The more expensive model also features two drawers as well as doors while the cheaper model did not include drawers. Both cabinets came with stools. The more expensive cabinet sold for $12.50 while the cheaper model was priced at $7.50. The nicer cabinet measured 32" high.

This South Bend Mfg. Co. cabinet probably dates from the late 1920s (before glass doors were added to the line of cabinets). It includes two doors on the top and two more on the bottom. The upper doors are decorated with parrot decals. The cabinet still includes its original label reading "South Bend Toys/Made by South Bend Manf. Co./South Bend, Ind." The cabinet measures 24" high x 16" wide x 10.25" deep. ($150-175).

Rose colored table and chairs, circa 1920s. The table measures 14" wide x 9.5" deep x 9.5" tall. The chairs are 12" tall x 7" wide x 7.5" deep. The table is decorated with a decal of a child. These chairs are short and wider than most chairs. Two wood Schoenhut dolls are having "tea" at the table. ($50 set).

Pictured is an example of the decorated tin tea sets produced by the Ohio Art Co., of Bryan, Ohio, for several decades. The early sets included this type of cup along with saucers, teapots, and sometimes a creamer and sugar, larger plates, and a tray. This design is by Fern Bisel Peat and dates from the early 1930s. Earlier designs include sets decorated with children's play scenes (circa 1917), Blue Birds (1923), Butterflies (1930), and Bo Peep (1931). The pictured set includes scenes from nursery rhymes, including Jack Be Nimble, Bo Peep, Tom the Piper's Son, and Mary Mary Quite Contrary. The Sears catalog for 1931 advertised a Bo Peep set with four cups, saucers, plates, and a pitcher and tray for 39 cents. (Place setting as shown $15-20).

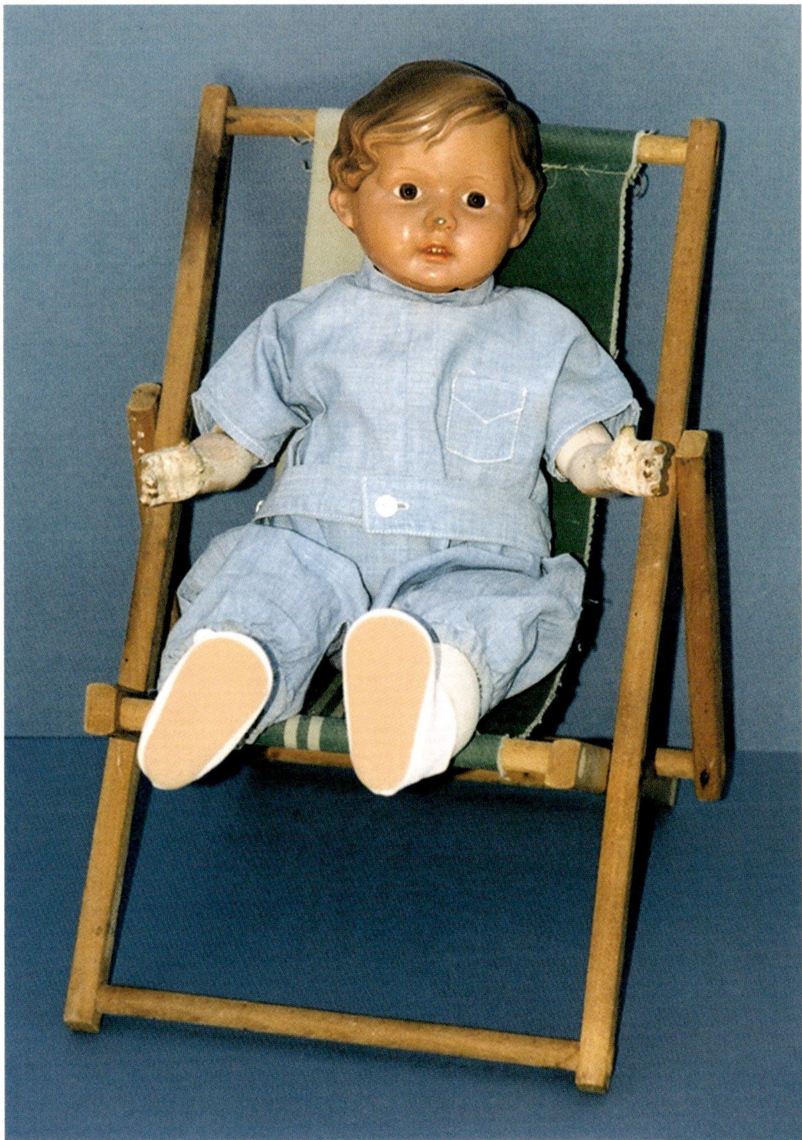

Oak child's rocker in the Mission style, circa 1920s. Children's rockers provide excellent seating space for the currently popular vintage and antique dolls in large sizes of 36" to 42". The chair measures 21" high x 13" deep (not including rockers) x 15" wide. ($75).

Folding beach chair made of wood and canvas, circa 1920s. It measures 16" high by 9" wide. A much smaller chair 6.75" tall by 2.25" wide was advertised in the Nerlich wholesale catalog for 1924. ($25 plus).

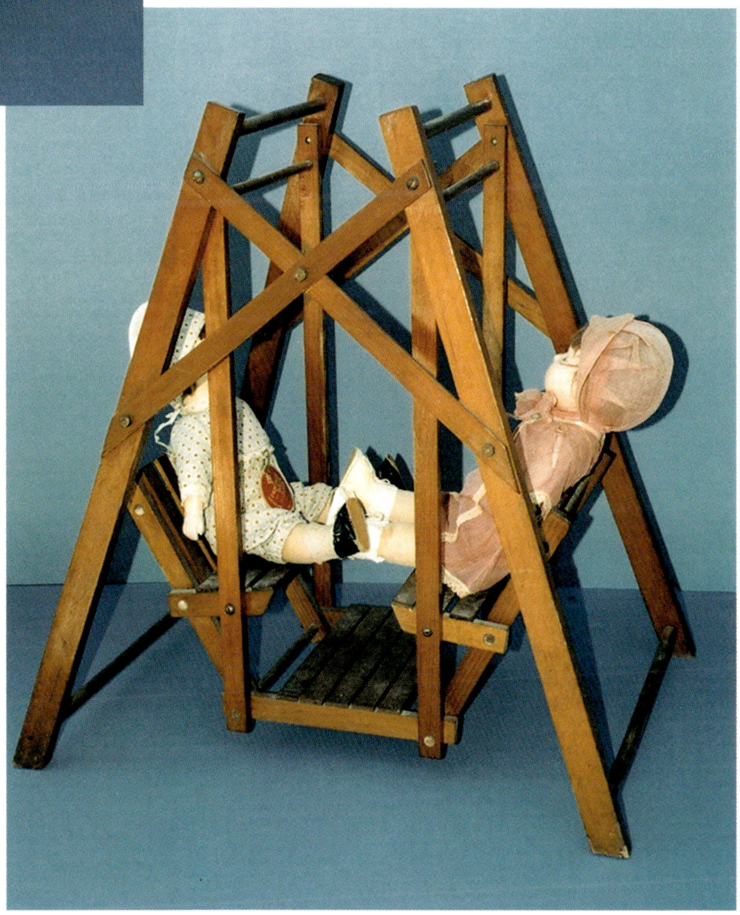

Doll swings of this type were made from the 1920s through the 1940s. This one is an older model. It measures 23" tall x 21.5" wide at its base. Two 12" dolls can sit comfortably in this swing. This model has a platform under the seats which some swings do not have. ($50).

Bibliography

Adams, Margaret. *Collectible Dolls and Accessories of the Twenties and Thirties from Sears, Roebuck & Co. Catalogues*. New York: Dover Publications, Inc., 1986.

Anderton, Johana Gast. *The Collector's Encyclopedia of Cloth Dolls*. Lombard, Illinois: Wallace-Homestead Book Co., 1984.

Anderton, Johana Gast. *Twentieth Century Dolls From Bisque to Vinyl*. North Kansas City, Missouri: Trojan Press, Inc.

Angione, Genevieve. *All Bisque and Half-Bisque Dolls*. Rochester, N.Y.: Thomas Nelson & Sons, 1969.

Axe, John. *The Encyclopedia of Celebrity Dolls*. Cumberland, Maryland: Hobby House Press, Inc., 1983

Barlow, Ronald, ed. *The Great American Antique Toy Bazaar*. El Cajon, California: Windmill Publishing Co., 1998.

Bucholz, Shirley. "Celluloid Dolls-Real Cinderellas." *Doll Reader*. October 1983, pp. 68-76.

Butler Brothers Catalogs. Chicago: Various Issues, 1920s.

Cobb, David. *Cobb's Doll Auction Catalog*. Columbus, Ohio: April 4, 2004.

Coleman, Dorothy, Elizabeth, and Evelyn. *The Collector's Encyclopedia of Dolls*. New York: Crown Publishers, 1968.

Coleman, Dorothy, Elizabeth, and Evelyn. *The Collector's Encyclopedia of Dolls, Volume II*. New York: Crown Publishers, 1986.

Coleman, Dorothy S. and Evelyn Jane. "Dolls During World War I." *Doll Reader*. October 1991, pp. 66-83.

Coleman, Dorothy, ed. *My Darling Dolls Kämmer & Reinhardt, 1927* (reprint of doll catalog). Des Moines, Iowa: Wallace-Homestead Book Co., 1972.

Coleman, Evelyn Jane and Dorothy. "Bébe Jumeau Dolls." *Doll Reader*. June-July 1993, pp.180-187.

Coleman, Evelyn Jane. "Poupées Raynal and Other French Molded Cloth-faced Dolls." *Doll Reader*. October 1989, pp. 110-119.

Coleman, Evelyn J., ed. *1914 Marshall Field & Co. Kringle Society Dolls*. Cumberland, Maryland: Hobby House Press, 1980.

Cook, Carolyn. "S.F.B.J. Character Dolls." *Doll Reader*. November 1983, pp. 74-80.

Cooper, Marlowe, Dee Van Kampen. *A Taste of Honey*. Escondido, California: Omni Publishers, 1976.

Corson, Carol. "Kamkins Babies Through Time." *Antique Doll Collector*. October 2002, pp. 30-34.

Corson, Carol. *Schoenhut Dolls*. Cumberland, Maryland: Hobby House Press, Inc., 1993.

DeFeo, Barbara, Carol Stover. *Doll Values Eighth Edition*. Paducah, Kentucky: Collector Books, 2004.

DeMiller, Suzanne L., and Dennis F. Brevik. *Arranbee Dolls*. Paducah, Kentucky: Collector Books, 2003.

Endo, Susan. "Pincushion Dolls by Dressel & Kister." *Doll Reader*. December 1985-January 1986, pp. 81-84.

F.A.O. Schwarz 1930 Christmas Catalog (reprint). Jeanne Breslin Vintage Catalogs. Evanston, Illinois: 2004.

Foulke, Jan. "All-Bisque Dolls by Kling & Co." *Doll Reader*. October 1990, pp. 167-169.

Foulke, Jan. *Kestner King of Dollmakers*. Grantsville, Maryland: Hobby House Press, Inc. 2000.

Foulke, Jan. "Raggedy Ann – Sweet All the Way Through." *Doll Reader*. September 1992, pp. 82-89.

Foulke, Jan. *16th Blue Book*. Grantsville, Maryland: Hobby House Press. 2003.

Foulke, Jan. "Steiff Dolls." *Doll Reader*. August-September 1986, pp. 106-117.

Foulke, Jan. "A Study of Käthe Kruse Doll I." *Doll Reader*. December 1991-January 1992, pp. 60-67.

Girardot, Susan. "The Five-in-One Doll." *Doll Reader*. September 1993, pp. 194-196.

Groninger, Margaret. "Comic Strip Pillow Dolls from the 1920s Period." *Doll Reader*. March 1980, pp. 18-19.

Hake, Ted. *Hake's Guide to Comic Character Collectibles*. Radnor, Pennsylvania: Wallace-Homestead Book Co. 1993.

Horn, Maurice, ed. *The World Encyclopedia of Comics*. New York: Chelsea House Publishers, 1976.

Horsman, E.I. *Babyland Rag Dolls*. Catalog circa 1912 (reprint). New York: E.I. Horsman.

Holderbaum, Linda Poirier. "Offerings From the Cereal City Part I." *Doll Reader*. August-September 1990, pp. 206-211.

Izen, Judith. *Collector's Guide to Ideal Dolls*. Paducah, Kentucky: Collector Books, 1994.

Jensen, Don. "Amberg's Silent Screen Stars, The First Movie Celebrity Dolls." *Antique Doll Collector*. August 2002, pp. 42-44.

Jensen, Don. *Collector's Guide to Horsman Dolls 1865-1950*. Paducah, Kentucky: Collector Books, 2002.

Kämmer & Reinhardt. *Mein Liebling Catalog* (reprint). North Hollywood, California: Pat Schoonmaker. Virginia A. Olsen owner of original catalog.

Kämmer & Reinhardt. *"My Darling Dolls" 1927 Catalog* (reprint). Dorothy S. Coleman introduction. Des Moines, Iowa: Wallace-Homestead Book Co., 1972.

Karl, Michele. *Composition and Wood Dolls and Toys*. Dubuque, Iowa: Antique Trader Books, 1998.

Lamp Post Antiques, ed. *Dolls Toys-Miniatures 1900-1930*. Colton, California: late 1960s, early 1970s.

Lipinski, Angelika. "Celluloid and the Doll Industry in Germany- 1896-1960." *Doll Reader*. September 1988, pp. 123-127.

Lipinski, Angelika. "Celluloid Dolls by Bushow & Beck." *Doll Reader*. September 1992, pp. 66-70.

Lipinski, Angelika. "The History of the Schildkrot Dolls." *Doll Reader*. January 1990, pp. 130-137. "Part II 1930-1960." February-March 1990, pp. 118-123.

Manos, Susan. *Schoenhut Dolls & Toys*. Paducah, Kentucky: Collector Books, 1976.

Marion, Frieda. "Half-Doll Collectors in England." *Doll Reader*. April 1986, pp. 106-111.

Mertz, Ursula R. *Collector's Encyclopedia of American Composition Dolls, Volume I*. Paducah, Kentucky: Collector Books, 1999.

Mertz, Ursula. *Collector's Encyclopedia of American Composition Dolls, Volume II*. Paducah, Kentucky: Collector Books, 2004.

Millman, Bernice. "A Precious Pair in a Collector's Ark." *Antique Doll Collector.* February 2003, pp. 34-36.

Montgomery Ward Catalog. Chicago: Various Issues, 1910-1930.

Moskovsky, Eva. "A Schlopsnies for the Steiff Firm." *Doll Reader.* November 1983, 70-73.

Murray, Lynn. "Käthe Kruse Then and Now." *Antique Doll Collector*. September 2002, pp. 21-26.

M.W. Savage Co. Fall-Winter Catalog. Minneapolis, Minnesota: 1929-1930.

Nicholson, Susan Brown. "Kate Jordan's Happifats." *Doll Reader.* October 1988, pp. 90-93.

Pennegues, Dominique. "French Celluloid Dolls, Part I Anel." *Doll Reader*. July 1991, pp. 98-102.

Pennegues, Dominique. "French Celluloid Dolls Part II Convert." *Doll Reader.* September 1991, pp. 98-103.

Pennegues, Dominique. "French Celluloid Dolls, Petitcollin." *Doll Reader*. October 1991, pp. 102-111.

Pennegues, Dominque. "French Celluloid Dolls Société. *Doll Reader*. November 1991, pp. 108-113.

Pennegues, Dominique. "French Celluloid Dolls Société Industrielle de Celluloid." *Doll Reader.* December 1991-January 1992, pp. 98-105.

Ragan, David. *Who's Who in Hollywood 1900-1976*. Rachelle, N. Y.: Arlington House Publishers, 1976.

Rait, Lila. "Children With Attitude, The Character Dolls of Kämmer and Reinhardt and J.D. Kestner." *Antique Doll Collector*. March 2002, pp. 28-31.

Robinson, Julie P. *Celluloid Dolls, Toys and Playthings*. Paducah, Kentucky: Collector Books, 2006.

Robison, Joleen Ashmon, and Kay Sellers. *Advertising Dolls: Identification & Value Guide*. Paducah, Kentucky: Collector Books, 1992.

St. George, Eleanor. *The Dolls of Yesterday*. New York: Charles Scribner's Sons, 1948.

Schoonmaker, Patricia. "Effanbee Dolls: The Formative Years, New/Old Discoveries." *Doll Reader*. February-March 1986, pp. 108-120.

Schroeder, Joseph, ed. *Toys, Games, and Dolls*. Northfield, Illinois.: Digest Books, Inc., 1971.

Sears, Roebuck and Company Catalogs. Chicago: Various Issues, 1911-1931.

Shufer, Jacque. "Bye-Lo Babies." *Antique Doll Collector*. December 2002, pp. 44-47.

Shafer, Jacque. "Ella Smith Doll Manufacturer." *Antique Doll Collector*. June 2002, pp. 45-47.

Siccardi, Rosita. "Rare Lenci Catalog Discovered." *Antique Doll Collector*. pp. 35-38.

Smith, Patricia, *Doll Values Fifth Series*. Paducah, Kentucky: Collector Books, 1987.

Smith, Patricia. *Doll Values Tenth Edition*. Paducah, Kentucky: Collector Books, 1994.

Spadaccini-Day, Barbara. "Doll Celebrities, Mainly French: Part I." *Doll Reader.* October 1988, pp. 142-149

Tagger, Marina. "Kämmer and Reinhardt Mold 100." *Antique Doll Collector*. February 2003, pp. 41-43

Theriault's Auction Catalog January 8, 1999. Annapolis, Maryland.

Theriault's Doll Auction Catalog. "Coming to America." September 2, 1995. Annapolis, Maryland.

Theriault, Florence. *Hertwig & Co. Archives. 1890-1937*. Annapolis, Maryland: Gold Horse Publishing, 2000.

Trotter, Gillian. *Norah Wellings: Cloth Dolls and Soft Toys*. Grantsville, Maryland: Hobby House Press, 2003.

Twigg, Bettyanne Bethea. "Cloth Dolls Part I." *Doll Reader*. November 1983, pp. 82-90.

VanLuven, Sally and Susan Graham. *Half-Dolls Price Guide*. Grantsville, Maryland: Hobby House Press, Inc., 2004.

Whitton, Margaret. "Raleigh Dolls." *Doll Reader*. February-March 1988, pp. 78-82.

Wlaschin, Ken. *The Illustrated Encyclopedia of the World's Great Movie Stars*. New York: Harmony Books, 1979.

Zillner, Dian. *Dolls and Accessories of the 1930s & 1940s*. Atglen, Pennsylvania: Schiffer Publishing, Ltd., 2002.

Zillner, Dian. *Hollywood Collectibles*. West Chester Pennsylvania: Schiffer Publishing Ltd., 1991.

Zillner, Dian. *Hollywood Collectibles The Sequel*. Atglen Pennsylvania: Schiffer Publishing Ltd., 1994.

Index

A

Aetna Doll and Toy Co., 68, 81
Alabama Indestructible Dolls, 50
Amberg, Louis & Son, 15, 25, 68-71, 120, 132, 135, 146, 151
- Am Kid, 68
- Bobby Blake & Dolly Drake, 120
- Bunny, John, 135
- Chaplin, Charlie, 135
- "It", 70
- Mibs, 69, 70
- Newborn Babe, 25, 68
- Sunny Orange Maid, 68
- Tiny Togs, 70
- Vanta Baby, 25, 71
- Victory Doll, 151
- Walking Doll, 69

American Character Doll Co., 71, 72, 148
- Bottletot, 71
- Happytot, 72
- Puggy, 72
- Sally, 72

Anel Co., 47
Angione, Genevieve, 15, 17
Arkadelphia Milling Co., 66
Arnold Print Works, 51
- Columbian Sailor, 51
- Little Red Riding Hood, 51
- Palmer Cox Brownies, 51
- Pickaninny, 51

Arranbee Doll Co., 91
- My Dream Baby, 91

Art Craft Toy Products, Inc., 146, 152
Art Fabric Mills, 51, 52, 123
Atlas Doll & Toy Co., 104-106
Aunt Jemima, 50, 66, 67
Averill, Georgene, 15, 24, 25, 49
- Bonnie Babe, 15, 24, 25

Averill Mfg. Corp. (Mme Hendren), 73-76
- Baby Brite, 76
- Dolly Record, 74
- Little Brother, Little Sister, 75
- Rock-a-Bye Baby, 74
- Sunny Girl (or boy), 75
- Whistling Rufus, 75

B

Baby BoKaye, 15, 24
Baby Bumps, 81
Babyland Rag, 52-54, 81
Babs Manufacturing Corp., 146
Baby Peggy (Montgomery), 15, 132, 134
Baker, Josephine, 139
Barney Google, 119, 120
BéBé Jumeau, 10
Beck Manufacturing Co., 146, 152
Bed, 162
Bergmann, C.M., 6
Berwick Doll Co., 91, 92
Bester Doll Manufacturing Co., 146, 152
Billiken, 81
Bissell's Carpet Sweeper, 161
Blondie, 119
Bobby Blake and Dolly Drake, 119, 120
Bonnet Head Doll, 7, 8, 27, 31
Bonnie Babe, 15, 24
Bringing Up Father, 119-121
Bronson, Betty, 132
Brownies, 119, 121
Bruckner, Albert, 52, 54, 55
- Dolly Pop, 54
- Tu-In-One, 54

Bubbles, 80
Buckerer, A., 99, 119, 129
Buggy, 166
Bunny, John, 132, 135
Buschow & Beck, 44, 46, 99, 101-103
Buster Brown, 66, 120, 121
Bye-Lo, 15, 23-26, 110

C

Cabinet, 167, 169
Cass, N.D., 160, 164
Cedar Chest, 159
Celluloid Novelty Co., 43
Century Doll Co., 24, 29, 146, 153
Ceresota Flour Boy, 66
Chad Valley Co., Ltd., 50, 60
Change-O-Doll Co., 91
Chaplin, Charlie, 132, 135
Chase, Martha, 55, 56
Chase Stockinet Dolls, 55, 56
Cream of Wheat (Rastus), 66
Chein, J.& Co., 106
Chicago Live Long Toys, 119, 124, 125
Cigarette Doll, 139, 140
Clear, Emma, 118
Coleman, Harry, 118
Colonial Toy Mfg. Co., 146, 153
Coogan, Jackie, 132, 136
Cox, Palmer, 121
Cradle, 158
Crib, 164

D

Daisy, 98 (composition), 7 (bisque)
Dean's Rag Book Co., 60, 138
Diana, 99
Dimples, 81, 85
Dishes, 156, 161, 170
Doll Carts, 159, 166
Dolly Dimples, 122
Dolly Record, 74
Dolly Walker Dolls, 118
Donahey, W., 15
Drayton, Grace, 68, 81, 119, 122
- Bobby Blake and Dolly Drake, 119, 120
- Campbell Kids, 82
- Dolly Dimples, 122
- Dolly Dingle, 122

Dressel & Kister, 141, 143
Dresser, 156, 157, 162, 164
Drukker, Hazel, 68
Mibs, 69, 70
Dy-o-la Dye Doll, 66

E

Effanbee, 68, 76-80, 119, 146, 154
- Baby Dainty, 68
- Baby Grumpy, 76, 78
- Bubbles, 80
- Harmonica Joe, 77
- Mary Jane, 154
- Rosemary, 79

Mae Starr Phonograph Doll, 79
Whistling Jim, 77
Elektra Toy & Novelty Co., 68, 146
Ella Cinders, 119, 122

F

Famlee Doll, 91
Felix the Cat, 119
Fiberoid Doll Products Co., 92
Fleischaker & Baum, 68, 76-80, 119, 146, 154
Foxy Grandpa, 119, 123
Freckles and His Friends, 119
Fulper Pottery Co., 24, 81, 141, 143, 146, 149

G

Gasoline Ally, 119, 124
Gene Carr Kids, 124
Gerling Toy Co., 60
Giebeler-Falk Doll Corp., 99, 104
Gladdie, 37
Goebel, F. & W., 141, 143
Goldberger, Eugene, 93
The Grumps, 119, 125
Grumpy, 76, 78

H

Handwerck, Heinrich, 7
Happifats, 15, 17
Happy Hooligan, 119
Harold Teen, 119
HEbee SHEbee, 119, 125
Heller, Alfred, 99
Hertal, Schwab & Co., 23, 26
Hertwig & Co., 7-9, 15, 16, 27, 119, 125, 131, 141, 143, 144
Heubach, Ernst, 9, 34
Heubach, Gebrüder, 15, 18, 24, 27, 34-37
Chin Chin, 15
Baby Stewart, 24, 27
Coquette, 34
Singing Girl, 35
Whistler, 36
High Chair, 158, 163
Horsman, E.I., 15, 24, 52, 54, 81-86, 120, 122, 124, 146, 149
Billiken, 68, 81
Baby Bumps, 68, 81
Baby Horsman, 85
Black Cotton Joe, 82
Campbell Kids, 68, 83
Candy Kid, 82
Coaster Kid, 83
Coogan, Jackie, 136
Dimples, 85
Dolly Rosebud, 85
Ella Cinders, 122
Peterkins, 84
SHEbee, 125
Hubley Manufacturing Co., 161, 167

I

Ice Box, 167
Ideal Novelty & Toy Co., 68, 86-90, 119, 130
Buster Brown, 120, 121
Farmer Boy, 87
Flossie Flirt, 90
Liberty Boy, 147
Mr. Hooligan, 68
Naughty Marietta, 68
Smiles, 90
Snookums, 130
Soozie Smiles, 89
Uneeda Biscuit Kid, 87
Yellow Kid, 86
I G – Farber Group, 44

J

Jaymar Specialty Co., 120, 128, 129
Jensen, Helen W., 37
Joe Palooka, 119
Johnny Gruelle, 50, 58
Jordan, Kate, 15, 17
Jumeau, 10
Juno, 44, 99, 100

K

K & K Toy Co., 27
Kallus, Joseph L., 15, 24
Kamkins Dolls, 57
Kämmer & Reinhardt, 24, 28, 38, 43, 44
Carl, 38
Elsa, 38
Elsie, 38
Gretchen, 38
Hans, 38
Marie, 38
Peter, 38
Walter, 38
Kampes, Louise R., 57
Käthe Kruse, 50, 61, 62
Katzenjammer Kids, 119
Kearley, V. M. Manf. Co., 168
Kewpies, 119, 126, 127
Kruse, Käthe, 50, 61, 62
Kelloggs, 50, 66, 67
Kestner, J. D., 11, 20, 21, 23-25, 28, 29, 39, 44
Hilda, 29
O.I.C. Baby, 29
Smiling Century Baby, 24, 29
"Wrestler", 20
Kling & Co., 22
Krazy Kat, 119

L

Lady Bountiful, 119, 124
Lenci, 50, 63-65, 140
Liberty Boy, 147
Limbach, 19
Lindberg, Charles, 132
Little Annie Rooney, 15, 119
Little Brother (or Sister), 75
Little Mary Mixup, 119
Little Nemo in Slumberland, 119
Little Orphan Annie, 119, 128
Lloyd, Harold, 132, 137

M

Mae Starr Phonograph Doll, 79
Manhattan Doll Co., 146-148
Marks Brothers, 48
Marseille, Armand, 11, 12, 24, 30, 31, 39, 40
Dream Baby, 30
Fany, 40
Flapper, 40
Googly Eyed, 39, 40
Just Me, 41
Kiddie Joy, 30
Queen Louise, 11
Mary Jane, 154
Max and Moritz, 129
Melville, Rose, 132
Metal Doll Co., 99
Metal Ware Corp., 168
Mibs, 15, 69, 70
Michtom, Morris, 86
Minerva, 44, 99, 101-103
Miss Flaked Rice, 66
Miss Korn Krisp, 66
Miss Phoebe Primm, 66
Moon Mullins, 120, 129
Morimura Brothers, 24, 33, 146, 150
Mutt & Jeff, 120, 129
Mutual Novelty Co., 138, 140

N

National Joint-Limb Doll Co., 146, 154
Natural Doll Co., 93
Nelke Corp., 57, 58
Newlyweds, 119, 130
Nippon, 19, 33
Nize Baby, 119, 130

O

O'Neill, Rose, 119, 126, 127
Ohio Art, 156, 170
Orsini, Jeanne I., 15, 23

Didi, 15
Mimi, 15, 23
Vivi, 15
Our Gang, 132, 137

P

Parlor Set, 156, 157
Parsons-Jackson Co., 48, 49
Peterkins, 84, 86
Peter Rabbit, 119
Pickford, Mary, 15, 61
Pinky Winky Products Co., 49
Petitcollin, 47
Petzold, Dora, 94
Popeye, 119
Putnum, Grace Storey, 15, 23-26

Q

Queue San Baby, 15, 17

R

Raggedy Ann, 50, 58
Raleigh, Jessie McCutcheon, 95
Raynal, 61
Recknagel, 13, 24, 32
Rheinische, Gummi and Celluloid Fabrik Co., 43-46
Rocker, 158, 163, 165, 171
Rollinson Doll, 59
Rollinson, Gertrude F., 59

S

S.& H. Novelty Co., 60
Scavini, Enrico and Elandi, 63-65
Schneider, Karl, 141, 143, 144
Schnickel-Fritz, 111
Schoberl & Becker, 44, 46
Schoenhut, A. & Co., 110-117, 119-121
Barney Google, 120
Maggie & Jiggs, 121
Schnickel-Fritz, 111, 120
September Morn, 15
S.F.B.J, 42
Shoofly, 163
Simon & Halbig, 24, 32, 41
Skippy, 119, 130
Smith, Ella, 50
Smitty, 119, 131
Snow Baby, 8
Societe Nobel Francaise, 47
South Bend Mfg.Co., 169
Standfuss, Karl, 44, 99, 100
Steiff, Margarete, 63
Stove, 128, 161, 167, 169
Swaine & Co., 24, 32
Lori, 24, 32
Sweet Petiekin, 58
Swing, 171

T

Table and Chairs, 157, 165, 170
Teenie Weenie, 15
Tillie the Toiler, 119
Toonerville Folks, 119
Toots and Casper, 119, 131
Trego Doll Mfg. Co., Inc., 146, 155
Trowbridge, Helen, 81
Trunk, 165
Tynie Baby, 15, 24, 81

U

Uncle Mose, 67
Unis France, 13
Utley Co., 59

V

Valentino, Rudolph, 132
Victory Doll, 151
Viscoloid Co., 48, 49
Volkstedt, Rudolstadt, 143

W

Webster, Joanne, 140
Weiss, Kuhnert & Co., 141, 142
Wellings, Norah, 50, 61, 138
Winnie Winkle, 119, 131
Wislizenus, Adolf, 42
Wolf, Louis & Co., 16
Wood Toy Co., 118

Z

Zast, P.R., 47